# The man who counted

# The man who counted

*a collection of mathematical adventures*

Malba Tahan

*Illustrated by Patricia Reid Baquero &*
*Translated by Leslie Clark and Alastair Reid*

W. W. Norton & Company New York London

First American edition 1993
Printed in the United States of America

The text of this book is composed in Perpetua
with the display set in Cochin
Composition and Manufacturing by The Maple-Vail Book Manufacturing Group
Book design by Guenet Abraham

Library of Congress Cataloging-in-Publication Data
Tahan, Malba, 1895–
[Homem que calculava. English]
The man who counted : a collection of mathematical
adventures / by Malba Tahan : translated by Leslie Clark and
Alastair Reid.
p. cm.
Translation of: O homem que calculava.
1. Mathematical recreations. I. Title.
QA95.T2913 1993
793.7′4—dc20 92-18822

ISBN 0-393-03430-5 (cloth)
ISBN 0-393-30934-7 (paper)

W. W. Norton & Company, Inc.
500 Fifth Avenue, New York, N.Y. 10110
www.wwnorton.com

W. W. Norton & Company Ltd.
Castle House, 75/76 Wells Street, London W1T 3QT

5 6 7 8 9 0

To the memory of seven great geometrists,
Christian or agnostic:

Descartes, Pascal, Newton
Leibniz, Euler, Lagrange, Comte

*Allah take pity on these infidels!*

and to the memory of the unforgettable mathematician, astronomer, and Muslim philosopher

Abu Jafar Muhammad ibn-Musa al-Khwarizmi

*Allah preserve him in his glory!*

and also to all who study, teach, or admire the prodigious science of scale, form, numbers, measures, functions, movement, and the laws of nature.

*I, pilgrim, descended from the Prophet*
*Ali Iezid Izz-Edim ibn-Salim Hanak*
*Malba Tahan*

*believer in Allah*
*and in Muhammad, his sacred Prophet*

dedicate these pages of legend and fantasy.

—Baghdad, nineteenth day of the moon of Ramadan, 1321

# Contents

# The man who counted

# 1

# A meeting of the minds

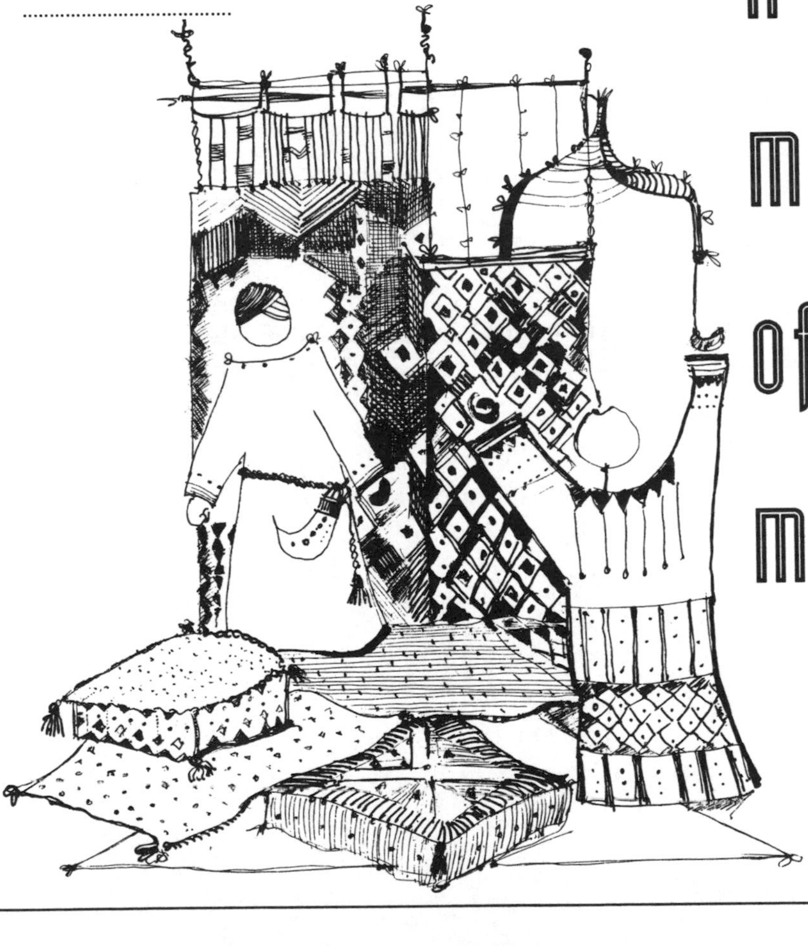

*Of the amusing circumstances of my encounter with a strange traveler on the road from Samarra to Baghdad.*

In the name of Allah, the All-Merciful!

My name is Hanak Tade Maia. Once I was returning, at my camel's slow pace, along the road to Baghdad after an excursion to the famous city of Samarra, on the banks of the Tigris, when I saw a modestly dressed traveler who was seated on a rock, apparently resting from the fatigue of the journey.

I was about to offer the perfunctory salaam of travelers when, to my great surprise, he rose and said ceremoniously, "One million, four hundred and twenty-three thousand, seven hundred and forty-five." He quickly sat down and lapsed into silence, his head resting in his hands, as if he were absorbed in profound meditation. I stopped at some distance and stood watching him, as if he were a historic monument to the legendary past.

A few moments later, the man again rose to his feet and, in a clear, deliberate voice, called out another, equally fabulous number, "Two million, three hundred and twenty-one thousand, eight hundred and sixty-six."

Several times more the strange traveler rose and uttered a number in the millions, before sinking down again on the rough stone by the roadside. Unable to restrain my curiosity, I approached the stranger and, after greeting him in the name of Allah, asked him the meaning of these fantastic sums.

"Stranger," replied the Man Who Counted, "I do not disapprove of this curiosity that disturbs the peace of my thoughts and calculations. And now that you have spoken to me with such courtesy and graciousness, I am going to accede to your wishes. But first I must tell you the story of my life."

And he told me the following, which, for your entertainment, I transcribe exactly as I heard it.

# 2

# Someone to count on

*In which Beremiz Samir, the Man Who Counted, tells the story of his life. How I learned of the brilliance of his calculations, and how we became traveling companions.*

.............................................................................

My name is Beremiz Samir. I was born in the little village of Khoi, in Persia, in the shadow of the huge pyramid of Mount Ararat. While still very young, I began work as a shepherd in the service of a rich gentleman from Khamat.

"Every day, at first light, I took the vast flock of sheep to graze and was required to bring them back to their fold before nightfall. For fear of losing a stray lamb and being severely punished as a consequence, I counted them several times a day.

"I became so good at counting that I could sometimes count the whole flock correctly at a glance. I went on to count, for practice, flights of birds in the sky. Little by little, I began to develop a great skill in this art. After a few months—thanks to new and continuing practice counting ants and other insects—I performed the remarkable feat of counting all the bees in a swarm. This prodigious calculation, however, was as nothing compared with the many others I later achieved. My generous master owned, in two or three far-off oases, huge date plantations, and, informed of my mathematical agility, he charged me with overseeing the sale of his fruit, which I counted in clusters, one by one. I worked thus, under the date palms, for almost ten years. Pleased with the profits I secured for him, my good master rewarded me with four months of rest, and I am now on my way to Baghdad to visit some of my family and to see the beautiful mosques and sumptuous palaces of the famous city. And, so as not to waste time, I have practiced throughout my journey counting the trees in this region, the flowers that perfume it, and the birds that fly among its clouds."

And, pointing to an old fig tree quite close, he went on, "That tree, for example, has two hundred and eighty-four branches. Given that each branch has, on the average, three hundred and forty-seven leaves, it is easy to conclude that that tree has a total of ninety-eight thousand, five hundred and forty-eight leaves. Well, my friend?"

"Wonderful!" I cried in astonishment. "It is incredible that a man can count, at a glance, all the branches in a tree, all the flowers in a garden. That skill can bring immense riches to anyone."

"Do you think so?" exclaimed Beremiz. "It has never occurred to me that counting millions of leaves and swarms of bees could make money. Who could possibly be interested in how many branches there are in a tree, how many birds in a flight that crosses the sky?"

"Your wondrous skill," I explained, "could be used in twenty thousand different ways. In a great capital like Constantinople, or even in Baghdad, you would be of invaluable help to the government. You could count populations, armies, and flocks. It would be easy for you to sum up the resources of the country, the value of its harvest, its taxes, its commodities, all the wealth of the state. Through my connections—for I am from Baghdad—I assure you that it will not be difficult to find some distinguished post in the service of Caliph al-Mutasim, our lord and master. Perhaps you might become treasurer, or fulfill the function of secretary to the Muslim household."

"If that is truly so, then my mind is made up," replied the counting man. "I am going to Baghdad."

And without more ado, he mounted behind me on my camel—the only one we had—and we set out on the long road to the splendid city. From that point on, united by that casual meeting on a country road, we became friends and inseparable companions.

Beremiz was a man of happy and talkative disposition. Still young (he was not yet twenty-six), he was blessed with a most lively intelligence and a remarkable aptitude for the science of numbers. From the most trivial of happenings, he would make unlikely analogies that demonstrated his mathematical acuity. He also knew how to tell stories and anecdotes that illustrated his conversation, already odd and attractive in itself.

At times, he would not speak for several hours, wrapped in an impenetrable silence, pondering prodigious calculations. On those occasions, I took pains not to disturb him. I left him in peace, to make, in his exceptional mind, fascinating discoveries in the arcane mysteries of mathematics, the science that the Arab race so developed and extended.

# 3

# Beasts of burden

*Of the singular episode of the thirty-five camels that were to be divided between three Arab brothers. How Beremiz Samir, the Man Who Counted, made an apparently impossible division that left the quarreling brothers completely satisfied. The unexpected profit that the transaction brought us.*

.....................................................................................

We had been traveling for a few hours without stopping when there occurred an episode worth retelling, wherein my companion Beremiz put to use his talents as an esteemed cultivator of algebra.

Close to an old, half-abandoned inn, we saw three men arguing heatedly beside a herd of camels. Amid the shouts and insults, the men gestured wildly in fierce debate, and we could hear their angry cries:

"It cannot be!"

"That is robbery!"

"But I do not agree!"

The intelligent Beremiz asked them why they were quarreling.

"We are brothers" the oldest explained, "and we received these 35 camels as our inheritance. According to the express wishes of my father, half of them belong to me, one third to my brother Hamed, and one-ninth to Harim, the youngest.

Nevertheless, we do not know how to make the division, and whatever one of us suggests, the other two dispute. Of the solutions tried so far, none have been acceptable. If half of 35 is 17½, if neither one-third nor one-ninth of this amount is a precise number, then how can we make the division?"

"Very simple," said the Man Who Counted. "I promise to make the division fairly, but let me add to the inheritance of 35 camels this splendid beast that brought us here at such an opportune moment."

At this point I intervened.

"But I cannot permit such madness. How are we going to continue on our journey if we are left without a camel?"

"Do not worry, my Baghdad friend," Beremiz said in a whisper. "I know exactly what I am doing. Give me your camel, and you will see what results."

And such was the tone of confidence in his voice that, without the slightest hesitation, I gave over my beautiful Jamal, which was then added to the number that had to be divided between the three brothers.

"My friends," he said, "I am going to make a fair and accurate division of the camels, which, as you can see, now number 36."

Turning to the eldest of the brothers, he spoke thus: "You would have received half of 35—that is, 17½. Now you will receive half of 36—that is, 18. You have nothing to complain about, because you gain by this division."

Turning to the second heir, he continued, "And you, Hamed, you would have received one-third of 35—that is, 11 and some. Now you will receive one-third of 36—that is, 12. You cannot protest, as you too gain by this division."

Finally, he spoke to the youngest: "And you, young Harim Namir, according to your father's last wishes, you were to receive one-ninth of 35, or 3 camels and part of another. Nevertheless, I will give you one-ninth of 36, or 4. You have benefited substantially and should be grateful to me for it."

And he concluded with the greatest confidence, "By this advantageous division, which has benefited everyone, 18 camels belong to the oldest, 12 to the next, and 4 to the youngest, which comes out to—18 + 12 + 4—34 camels. Of the 36 camels, therefore, there are 2 extra. One, as you know, belongs to my friend from Baghdad. The other rightly belongs to me for having resolved the complicated problem of the inheritance to everyone's satisfaction."

"Stranger, you are a most intelligent man," exclaimed the oldest of the three brothers, "and we accept your solution with the confidence that it was achieved with justice and equity."

The clever Beremiz, the Man Who Counted, took possession of one of the finest animals in the herd and, handing me the reins of my own animal, said, "Now, dear friend, you can continue the journey on your camel, comfortable and content. I have one of my own to carry me."

And we traveled on toward Baghdad.

# 4

# Food for thought

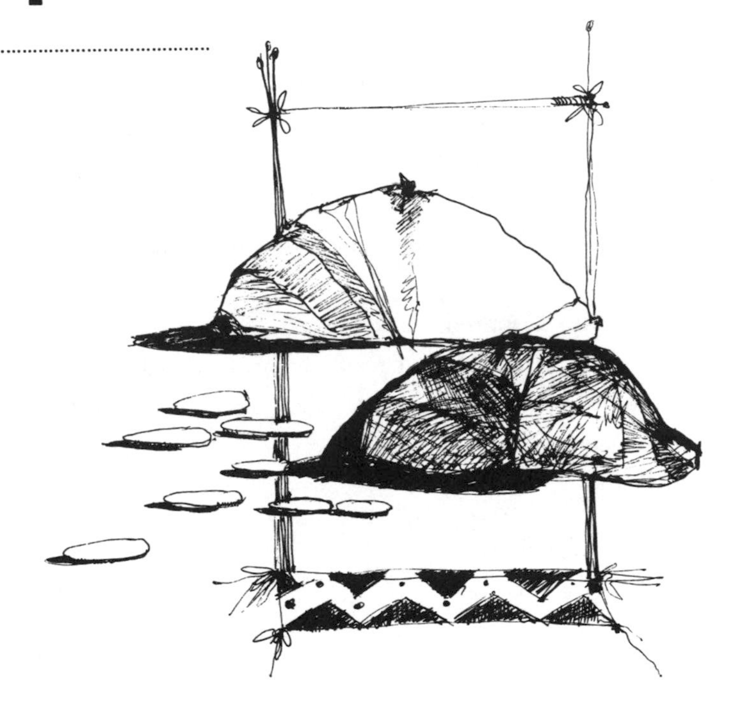

*Of our coming across a rich sheik, wounded and hungry. The offer he made to us for our eight loaves of bread, and how the division of the eight coins we received in payment was resolved in a surprising manner. Beremiz's three types of division: simple division, exact division, and perfect division. Praise for the Man Who Counted from an illustrious vizier.*

Three days later, we were approaching the ruins of a small village called Sippar, when we found sprawled on the ground a poor traveler, his clothes in rags and he apparently badly hurt. His condition was pitiful. We went to the aid of the unfortunate man, and he later told us the story of his misfortune.

His name was Salem Nasair, and he was one of the richest merchants in Baghdad. On the way back from Basra, a few days before, bound for el-Hilleh, his large caravan had been attacked and looted by a band of Persian desert nomads, and almost everyone had perished at their hands. He, the head, managed to escape miraculously by hiding in the sand among the bodies of his slaves.

When he had finished his tale of woe, he asked us in a trembling voice, "Do you by some chance have anything to eat? I am dying of hunger."

"I have three loaves of bread," I answered.

"I have five," said the Man Who Counted.

"Very well," answered the sheik. "I beg you to share those loaves with me. Let me make an equitable arrangement. I promise to pay for the bread with eight pieces of gold, when I get to Baghdad."

So we did.

The following day, in late afternoon, we entered the famous city of Baghdad, Pearl of the East. Crossing the bustling square, we were held up by a resplendent entourage, at whose head, on an elegant sorrel horse, rode the powerful Ibrahim Maluf, one of the viziers. Seeing Sheik Salem Nasair in our company, he brought his dazzling retinue to a halt and called out to him, "What happened to you, my friend? How is it that you arrive in Baghdad in rags, in the company of these two strangers?"

The poor sheik told him in detail everything that had happened on his journey, praising us effusively.

"Pay these two strangers at once," ordered the vizier. Taking from his purse eight gold coins, he gave them to Salem Nasair, saying, "I will take you to the palace at once, since the defender of the faithful will unquestionably want to be informed of this new affront by bandits and Bedouins, attacking our friends and sacking one of our caravans in the caliph's territories."

Then Salem Nassair said to us, "I take leave of you, my friends. I wish, however, to thank you once more for your help and, as promised, to repay your generosity."

Turning to the Man Who Counted, he said, "Here are five gold pieces for your five loaves."

Then to me, "And three to you, my Baghdad friend, for your three."

To my great surprise, the Man Who Counted made a respectful objection. "Forgive me, O Sheik! Such a division, although apparently simple, is not mathematically correct. Since I gave five loaves, I should receive seven coins. My friend, who supplied three loaves, should receive only one."

"In the name of Muhammad!" exclaimed the vizier, showing a lively interest. "How can this stranger justify such an absurd division?"

The Man Who Counted approached the minister and spoke as follows:

"Let me show you, O Vizier, that my proposal is mathematically correct. During the journey, when we were hungry, I took out a loaf and divided it into three pieces. We each ate one piece. My five loaves, then, yielded up fifteen pieces, correct? My friend's three loaves added nine pieces, making a total of twenty-four pieces. Of my fifteen pieces, I consumed eight, so that actually I contributed seven. Of my friend's nine pieces, he also consumed eight, so contributing one. My seven pieces and my friend's one made up the eight that went to Sheik Salem Nasair. So it is right that I receive seven coins and my friend only one."

The grand vizier, after praising the Man Who Counted extravagantly, ordered that he be given seven coins and I one. The mathematician's proof was logical, perfect, and irrefutable.

But, however just the division, it clearly did not satisfy Beremiz, who, turning to the surprised minister, went on, "This division, seven for me and one for my friend, is, as I just proved, mathematically perfect; but is is not perfect in the eyes of the Almighty."

And, gathering the coins again, he divided them equally, handing me four of them and keeping four.

“This is an extraordinary man,” declared the vizier. “He did not accept the proposed division of the eight coins into five and three. He then proved that he had a right to seven and his companion to one. But then he divides the coins into two equal parts and gives one of them to his friend.”

He added with enthusiasm, “By the Almighty! This young man, besides being wise and quick in the ways of arithmetic, is a fine and generous friend. He shall be my secretary this very day.”

“Great Vizier,” said the Man Who Counted. “I notice that you have just expressed, in thirty words and 125 letters, the highest praise I have ever heard. May Allah eternally bless and protect you!”

My friend Beremiz’s skill went as far as the words and letters being used. We all of us marveled at that display of enviable genius.

# 5

# In so many words

*Of the prodigious calculations Beremiz Samir performed, as we made our way to the Golden Goose Inn, in order to determine the exact number of words spoken in the course of our journey and the average number of words per minute. How the Man Who Counted resolves a problem.*

.......................................................................

After leaving Sheik Nasair and Vizier Maluf, we went along to a small inn called the Golden Goose, in the neighborhood of the Mosque of Suleiman. There we sold our camels to a camel driver of long acquaintance who lived nearby.

On the way, I said to Beremiz, "Now you see, my friend, that I was right when I said that a master of calculation with your talents could easily find a good job in Baghdad. As soon as you arrived, they asked you to accept the post of secretary to a vizier. Now you do not have to return to the sad, stony village of Khoi."

"Although I may prosper and become rich here," the counting man answered, "I want to return to Persia someday in order to see my native land again. Ungrateful is he who would forget his country and childhood friends when he finds happiness in an oasis of prosperity and good fortune."

And, taking my arm, he added, "We have traveled together for eight days exactly. During that time, clarifying points and mulling over things that interest me, I have spoken exactly 414,720 words. Since in eight days there are 11,520 minutes, one can deduce that during the journey I uttered an average of 36 words per minute—that is, 2,160 per hour. These numbers show that I spoke little, that I was discreet and did not waste your time with pointless discourses. A man of few words, excessively silent, becomes a disagreeable creature; but those who talk endlessly bore and annoy their companions. Therefore, we should avoid fruitless chatter, without becoming terse and, therefore, ungracious. To that end, I will tell you a very curious case."

After a brief pause, the counting man recounted the following:

"In Teheran, in Persia, there was an old merchant who had three sons. One day the merchant called his children together and said, 'He who can go a day without speaking one pointless word, I will reward with a prize of twenty-three gold coins.'

"At nightfall three sons presented themselves before their ancient parent. Said the first, 'Today, my father, I avoided all pointless words. I hope, therefore, that I have earned the promised reward. The prize, as you will recall, amounted to twenty-three pieces.'

"The second son approached the old man, kissed his hands and said no more than 'Good evening, Father.'

"The youngest spoke not a word. He approached his father and extended his open hand to receive the prize. The merchant, after observing the behavior of his three sons, said, 'The oldest, on coming before me, lost my attention with various pointless words; the youngest showed himself too terse. The prize, then, goes to my second son, who was circumspect but not verbose, simple, without pretension.' "

When he finished, Beremiz asked me, "Do you not think that the old man judged his three sons fairly?"

I did not answer. I thought it better not to discuss the twenty-three coins with that amazing man who was always reducing everything to numbers, calculating averages, and solving problems.

A few moments later we arrived at the Golden Goose. The innkeeper was named Salim and had once worked for my father. When he saw me, he grinned and cried, "Allah be with you, little one. I await your wishes now and forever."

I told him that I needed a room for myself and for my friend Beremiz Samir, the calculator, secretary to the vizier Maluf.

"This man is a mathematician?" asked old Salim. "Then he has come at just the right moment to help me out of a difficult spot. I have just had a dispute with a jeweler. We argued for a long time, and yet we still have a problem that we cannot resolve."

Having been informed that a great mathematician had arrived at the inn, several curious people drew round. The jeweler was summoned, and he declared himself extremely interested in a solution to the problem.

"What is the cause of the argument?" Beremiz asked.

"This man," old Salim said, pointing to the jeweler, "came from Syria to sell precious stones in Baghdad. He promised he would pay 20 dinars for his lodgings if he sold all of his jewels for 100 dinars, and 35 dinars if he sold them for 200. After several days of wandering about, he ended up selling all of them for 140 dinars. How much does he owe me according to our agreement?"

"24½ dinars! It stands to reason!" answered the Syrian. "If after selling the jewels for 200 dinars, I owed you 35, then, logically, if I had sold them for 20 dinars—ten times less—I would owe only 3½ dinars. But, as you well know, I sold them for 140 dinars, and 20 goes into 140 7 times, if my calculation is correct. Therefore, if I had sold the jewels for 20 and would have owed you 3½, having then sold them for 140, I now owe you an amount equal to 7 times 3½ dinars, or 24½ dinars.

PROPORTION SUGGESTED BY THE JEWELER

$$200 : 35 :: 140 : x$$

$$x = \frac{35 \times 140}{200} = 24\tfrac{1}{2}$$

"You are wrong," old Salim replied in irritation. "According to my count, it is 28. Listen! If for a sale price of 100 I was to receive 20, then for 140 I should get 28. It's obvious! I will show you."

Old Salim reasoned this way: "If for 100 dinars, I was to receive 20, then for 10—which is one-tenth of 100—I would receive one-tenth of 20, that is, 2. Therefore, for 10 I would receive 2. How many times does 10 go into 140? Fourteen. Therefore, for a sale of 140 I should receive 14 times 2, which is 28, as I already said."

Proportion Suggested by Old Salim

$$100 : 20 :: 140 : x$$

$$x = \frac{20 \times 140}{100} = 28$$

Old Salim, after all these calculations, exclaimed energetically, "I should receive 28! That is the correct amount!"

"Calm down, my friends," broke in the counting man. "We should settle matters quietly, with some decorum. Haste leads to anger and mistakes. The solutions you suggest are wrong, as I will show."

"According to the agreement that the two of you made, you," he said, speaking to the Syrian, "were to pay 20 dinars for your lodgings if you had sold the jewels for 100 dinars, but 35 if you had sold them for 200 dinars. Therefore, we have the following:

| *Sale Price* | *Price of Lodgings* |
|---|---|
| 200 | 35 |
| −100 | −20 |
| 100 | 15 |

"You notice that a difference of 100 dinars in the sale price corresponds to a difference of 15 in the price of lodgings. Is that clear?"

"As clear as camel's milk," agreed the two men.

"Then," the mathematician went on, "if an increase of 100 in the sale price means a difference of 15 in the lodgings, I ask you, 'What would be the increase in lodgings if the price increased by only 40?' If the difference were 20, which is one-fifth of 100, the increase in lodgings would be 3, since 3 is one-fifth of 15. For a difference of 40, which is double 20, the increase in lodgings would be 6. Therefore, after selling the jewels for 140, the bill would be 26 dinars.

PROPORTION SUGGESTED BY BEREMIZ

$$100 : 15 :: 40 : x$$

$$x = \frac{15 \times 40}{100} = 6$$

"My friends, numbers, in their utter simplicity, dazzle even the wisest of men. Even divisions that seem to us perfect are, at times, beset with error. From the uncertainty of calculations comes the undeniable prestige of the mathematician. According to the terms of the agreement, the man will have to pay 26 dinars and not 24½, as he at first thought. Even in the final solution to the problem, there was a small difference that should not be taken lightly, the size of which I cannot express in numbers alone."

"The gentleman is right," agreed the jeweler. "I see that my calculation was wrong."

And, without hesitating, he took 26 dinars from his bag and gave them to old Salim, at the same time offering the clever Beremiz a beautiful gold ring with dark stones, along with expressions of affection. All those gathered in the inn admired the wisdom of the counting man.

# 6

# Trial by numbers

*What happened during our visit to the Vizier Maluf. Our meeting with the poet who did not believe in the wonders of calculation. The Man Who Counted demonstrates an original way of counting the camels in an extensive caravan. The age of the betrothed and a camel's ear. Beremiz discovers "quadratic friendship" and speaks of King Solomon.*

After the second prayers, we left the Golden Goose and hurried to the house of Vizier Ibrahim Maluf, the king's minister. Entering his property, I was truly amazed.

We passed through the heavy iron gate and down a narrow corridor, led by a huge black slave with gold armbands, into the splendid inner garden of the palace. This garden, laid out with exquisite taste, was shaded by two rows of orange trees. Various doors opened on to it, some of which must have led to the harem. At the sight of us, two infidel slaves who were picking flowers fled among the flower beds and disappeared behind the columns. From the elegant garden, through a narrow door in a high wall, we crossed to a patio. In its center, splendidly tiled, stood a fountain with three water jets, three liquid curves sparkling in the sun.

Still following the slave with gold armbands, we crossed the patio and went into the palace itself. Passing through a variety of richly furnished rooms, hung with tapestries fringed in silver thread, we at last reached the rooms of the king's minister, who was reclining on large cushions and conversing with two friends.

One of them I recognized as Sheik Salem Nasair, our companion in the desert adventure. The other was a small, round-faced man, with a kindly expression, and a trace of gray in his beard. Finely dressed, he wore a rectangular medallion, one-half of which was golden yellow and the other half dark, like bronze.

Vizier Maluf received us with obvious goodwill and, turning to the man with the medal, said with a smile, "My dear poet, here is our great calculator. The young man with him is a citizen of Baghdad whom he met by chance while wandering in the ways of Allah."

We both gave a deep salaam to the noble sheik. We learned later that his companion was the famous poet Abdul Hajmid, close friend of Caliph al-Mutasim. The curious medal he had received from the hands of the caliph, as a prize for writing a poem of 30,200 verses without once using the letters kaf, lam, or ayn.

"I have some trouble believing, friend Maluf, in the prodigious feats of this Persian calculator," laughed the poet. "When numbers are combined, a certain trickery, an algebraic subtlety, comes into play. Once there appeared before King el-Harit, son of Modad, a wise man who declared he could read destinies in the sand. 'Do you make exact calculations?' the king asked him; and before the magus could recover from his surprise, he nodded, 'If you do not know how to calculate exactly, your visions are worth nothing; if you arrive at them through calculation alone, I disbelieve them.' In India, I learned a proverb that says, 'Distrust the calculation seven times over, the mathematician a hundred times.' "

"To quiet this distrust," suggested the vizier, "let us put our guest to a decisive test."

So saying, he rose from his cushions and, taking Beremiz lightly by the arm, led him to one of the balconies of the palace. He opened the casement onto yet another patio, which at that moment was full of camels, fine animals, almost all of them of a good breed. I noticed two or three white ones, from Mongolia, and a few clear-skinned *carehs*.

"Here is a fine collection of camels I bought yesterday," said the vizier. "I want to send them as a gift to the father of my betrothed. I know exactly how many there are. Can you tell me?"

And the vizier, to make the test more interesting, whispered the total to his friend. I was taken aback. There were a great many camels, constantly milling about. If my friend were to make a mistake, our visit would be a woeful disaster.

But after running his eyes over the restless herd, Beremiz said, "According to my calculations, O Vizier, there are 257 camels in this patio."

"Exactly! Quite correct!" the vizier confirmed. "Two hundred and fifty-seven camels, by Allah!"

"How did you manage to count them so swiftly and so exactly?" asked the poet, beside himself with curiosity.

"Simple," explained Beremiz. "Counting the camels one by one would be to me an uninteresting chore. So I went about it in the following way: I counted first the hooves and then the ears. They came to a total of 1,541. To this, I added 1 and then divided the result by 6. The exact quotient: 257."

"Great heavens!" exclaimed the vizier, in delight. "How original! Who could have imagined that he would count the ears and hooves just for interest's sake?"

"I should say, Vizier," Beremiz added, "that calculations are sometimes made difficult by carelessness or inability on the calculator's part. Once, in Khoi, when I was watching my master's flock, a cloud of butterflies passed. Another shepherd asked whether I could count them. 'Eight hundred and fifty-six,' I answered. 'Eight hundred and fifty-six?' he exclaimed, as if he found the total exaggerated. Only then did I realize that I had counted not the butterflies but their wings. I divided by two and got the correct total."

At this the vizier laughed aloud, a sound to my ears like sweet music.

"There's one point in all this that is beyond me," said the poet, very seriously. "I understand that dividing by 6—4 legs and 2 ears—would yield the total number of camels. But I do not understand why he added 1 before dividing his total of 1,541."

"Quite simple," replied Beremiz. "In counting the ears, I noticed that one of the camels had a slight defect: it lacked an ear. So, to round out the total, I had to add 1."

Then, turning to the vizier, he asked. "Would it be indiscreet or imprudent to inquire, O Vizier, how old your betrothed might be?"

"Not at all," replied the minister, smiling. "Astir is sixteen." And he added, with a trace of suspicion, "But I do not see any relation between her age and the camels I am about to present to my future father-in-law."

"I only wanted to make a small suggestion," replied Beremiz. "If you were to remove from the herd the defective animal, the total would be 256, which is the square of 16: 16 times 16. The gift offered to the father of the delightful Astir would then have a mathematical perfection, the number of camels equal to the square of the age of the beloved. The number 256 is an exact power of 2—a number held to be symbolic by the ancients—while 257 is a prime number. These relations between squared numbers are good auguries for lovers. There is an interesting legend about squared numbers. Would you like to hear it?"

"With pleasure," replied the vizier. "Good stories well told are a pleasure to listen to, and I am always eager to hear them."

Flattered, the Man Who Counted inclined his head graciously and began, "It is told of King Solomon that, as proof of his courtesy and his wisdom, he gave to his betrothed, the beautiful Belkis, queen of Sheba, a box containing 529 pearls. Why 529? Because 529 is the square of 23—that is, 23 times 23 gives 529—and 23 was the queen's age. In young Astir's case, 256 has the advantage over 529."

Everyone looked at the Man Who Counted in some astonishment. Serenely, he went on, "The digits of the number 256 add up to 13. The square of 13 is 169. The digits of 169 add up to 16. As a result, the numbers 13 and 16 have a curious relation, which we could call a quadratic friendship. If numbers could speak, we might overhear the following dialogue: Sixteen says to Thirteen, 'I wish to offer you an homage to our friendship. My square is 256, and the sum of its digits is 13.' And Thirteen would reply, 'Thank you for your kindness, dear friend. I wish to answer in the same coin. *My* square is 169, and the sum of *its* digits is 16.' I think that I have amply justified the preference we should grant to the number 256, which is much more singular than 257."

"Your idea is a rare one," replied the vizier. "I shall do it, although I could be accused of plagiarizing the great Solomon." And, turning to the poet, he said, "I see that this man's intelligence is no less than his skill at finding analogies and inventing stories. I did well when I decided to make him my secretary."

"I regret to have to inform you, illustrious lord," replied Beremiz, "that I can accept your noble offer only if there is also a place for my friend Hanak Tadé Maia, who is now without work or funds."

I was astonished and delighted by the graciousness of the Man Who Counted, who sought for me in this way the worthy protection of the powerful vizier.

"Your request is a fair one," replied the vizier. "Your friend may undertake the work of scribe, with its corresponding salary."

I accepted promptly and then expressed my gratitude to the vizier and to the good Beremiz.

# 7

# Going to market

*Of our visit to the marketplace. Beremiz and the blue turban. The case of "the four fours." The problem of the fifty dinars. Beremiz solves the problem and receives a very beautiful gift.*

Several days later, having finished our day's work in the vizier's palace, we went for a stroll in the marketplace and gardens of Baghdad. There was an unusually intense activity that afternoon because, only a few hours earlier, the rich caravans had arrived from Damascus. The arrival of the caravans was always a great event. It offered the only way to find out what other countries were producing and to mingle with the foreign tradesmen. The city was unusually active, bustling with life.

It was impossible to enter the shoemaker's bazaar, for example; there were sacks and boxes of new merchandise in all the patios and storerooms. Foreigners from Damascus, wearing huge and multicolored turbans and sporting weapons at their belts, strolled casually through the market, gazing indifferently at the stalls. There was a pronounced odor of incense, kif, and spice. The vegetable merchants were quarreling, nearly coming to blows and hurling insults at one another.

A young guitarist from Mosul, sitting on some sacks, was singing a sad, droning tune.

*What matters the life of a man*
*If people, for better or worse,*
*Live as simply as they can.*
*Here ends my verse.*

The shopkeepers, in the doors of their shops, were hawking their wares, making exaggerated claims with the fertile imagination of Arabs.

"This cloth! Look at it. Worthy of an emir!"

"Friends! Here is a delicious perfume that recalls the love of your wife."

"Look, O master, at these slippers and this beautiful caftan that the genies recommend to the angels!"

Beremiz was interested in an elegant, bright blue turban that a hunchbacked Syrian was selling for four dinars. This merchant's shop was quite unusual as well, for everything in it—turbans, boxes, daggers, bracelets, and so on—cost only four dinars. There was a sign that said in bright letters:

The Four Fours

Seeing that Beremiz was interested in buying the turban, I said, "Such an extravagance seems mad to me. We have only a little money, and we have not yet paid for our lodgings."

"It is not the turban that interests me," replied Beremiz. "Did you notice that this shop is called The Four Fours. This is a coincidence of unusual importance."

"A coincidence? Why?"

"The name of this business recalls one of the wonders of calculus: using four fours, we can get any number whatsoever."

And before I could ask him about this mystery, Beremiz explained by writing in the fine sand that was scattered on the floor.

"Do you want to get zero? Nothing simpler. Just write:

$$44 - 44$$

Here you have four fours in an equation that equals zero.

"What about the number 1? Here is the easiest way:

$$\frac{44}{44}$$

This fraction represents the quotient of 44 divided by 44, or 1.

"Would you like to see the number 2? We can easily use the four fours and write:

$$\frac{4}{4} + \frac{4}{4}$$

The sum of the two fractions is exactly 2. Three is even easier. Just write the expression

$$\frac{4+4+4}{4}$$

Notice that the dividend adds up to 12, which when divided by 4 yields 3. So 3, too, can be made from four fours."

"And how are you going to get the number 4?" I asked him.

"Nothing could be easier," Beremiz explained. "We can get 4 in several different ways. Here I have one expression equal to 4:

$$4+\frac{4-4}{4}$$

You can see that the right-hand part equals nothing, and that the total is 4. The expression is equivalent to $4+0$, that is, 4."

I saw that the Syrian merchant was listening carefully to Beremiz's explanation, as if the combinations of four fours fascinated him.

Beremiz continued, "If I want to get the number 5—no problem. We write:

$$\frac{(4\times 4)+4}{4}$$

This fraction shows 20 divided by 4, and the quotient is 5. Here we have 5 written out with four fours.

"Now we go on to 6, and here is a most elegant formula:

$$\frac{4+4}{4}+4$$

"A slight alteration in that formula gives us 7:

$$\frac{44}{4}-4$$

"To get 8 by means of four fours is easy:

$$4+4+4-4$$

"And the number 9 is also interesting:

$$4+4+\frac{4}{4}$$

"And now I will show you a beautiful expression,

$$\frac{44-4}{4}$$

equaling 10, made with four fours."

At that point, the hunchbacked shopkeeper, who had been following Beremiz's explanations in respectful silence, interrupted. "From what I have just heard, the gentleman must be a distinguished mathematician. If he can solve a certain riddle that I came across in a mathematical problem two years ago, I will give him the blue turban that he wanted to buy."

And the shopkeeper told the following: "Once I lent 100 dinars, 50 to a sheik from Medina and another 50 to a merchant from Cairo.

"The sheik paid the debt in four installments, in the following amounts: 20, 15, 10, and 5, that is

| | | | | | |
|---|---|---|---|---|---|
| *Paid* | *20* | *and* | *still* | *owed* | *30* |
| " | *15* | " | " | " | *15* |
| " | *10* | " | " | " | *5* |
| " | *5* | " | " | " | *0* |
| *Total* | *50* | | | *Total* | *50* |

"Notice, my friend, that the total of the payments and the total of his debt balance were both 50.

"The merchant from Cairo also paid the debt of 50 dinars in four installments, in the following amounts:

| | | | | | |
|---|---|---|---|---|---|
| *Paid* | *20* | *and* | *still* | *owed* | *30* |
| " | *18* | " | " | " | *12* |
| " | *3* | " | " | " | *9* |
| " | *9* | " | " | " | *0* |
| *Total* | *50* | | | *Total* | *51* |

"Note that the first total is 50—as in the previous case—while the other total is 51. Apparently this should not have occurred. I do not know how to explain the difference of 1 in the second manner of repayment. I know that I was not cheated, as I was paid all of the debt, but how to explain the difference between the total of 51 in the second case and 50 in the first?"

"My friend," began Beremiz, "I can explain this in a few words. The debt balance has nothing to do with the total debt. Let us agree that the debt of 50 was paid in three installments—the first, 10; the second, 5; and the third, 35. The bill, with the balance, would be

| | | | | | |
|---|---|---|---|---|---|
| *Paid* | *10* | *and* | *still* | *owed* | *40* |
| " | *5* | " | " | " | *35* |
| " | *35* | " | " | " | *0* |
| *Total* | *50* | | | *Total* | *75* |

In this example, the first total is still 50, while the total of the debt balance is, as you can see, 75; it could be 80, 99, 100, 260, 800, or any number at all. Only by chance would the total be exactly 50, as in the case of the sheik, or 51, as in the case of the merchant."

The shopkeeper was satisfied now that he understood Beremiz's explanation, and he kept his promise by giving the Man Who Counted the blue turban worth 4 dinars.

# 8

# Seventh heaven

*Beremiz holds forth on the forms of geometry. Our happy encounter with Sheik Salem Nasair and his friends the sheep rearers. Beremiz solves the problem of the twenty-one wine casks. The explanation of the disappearing dinar.*

Beremiz was very satisfied at receiving the fine present from the Syrian merchant. "It is very well made," he said, turning the turban over and looking at it from all sides. "It has, however, one defect, which could easily be avoided: its form is not strictly geometric."

I looked at him, unable to conceal my surprise. That original man had a way of transforming the most ordinary things, to the point of considering even turbans in the light of geometric form.

"It should not be surprising to you, my friend," said the wise Persian, "that I should wish for turbans of geometric form. *Geometry is everywhere.* Consider the ordinary and perfect forms of many bodies. Flowers, leaves, and innumerable animals reveal admirable symmetries that lighten the spirit. Geometry, I repeat, exists everywhere: in the sun's disk, in leaves, in the rainbow, in butterflies, in diamonds, in starfish, in the tiniest grain of sand. There is an infinite variety of geometric forms throughout nature. A crow flying slowly through the air traces wondrous figures with its sooty body. The blood circulating in the veins of a camel also obeys strict geometric principles; its humps, unique among mammals, show a singular elliptical form; the stone thrown at an intruding jackal describes a perfect curve in the air, known as a parabola: the bee makes its cells in the form of hexagonal prisms and uses that geometric form to build its house with the greatest possible economy of material.

"*Geometry exists everywhere.* It is necessary, however, to have eyes to see it, intelligence to understand it, and spirit to wonder at it. The rude Bedouin sees geometric forms but does not understand them; the Sunni understands them but does not admire them; the artist, finally, perceives the perfection of figures, understands beauty, and admires order and harmony. God was the Great Geometer. He geometrized heaven and hearth. In Persia there exists one plant, much sought as food by camels and sheep, whose seed . . ."

And so, holding forth enthusiastically on the multiple beauties embraced by geometry, Beremiz walked along the long and dusty road from the Place of the Merchants to the Bridge of Victory. I accompanied him in silence, enchanted by his curious enlightenments.

After crossing Muazen Square, which is also known as the Shelter of the Camel Drivers, we descried the beautiful Inn of the Seven Sorrows, much frequented in the hot weather by Bedouins and travelers from Damascus and Mosul. Its most elegant feature was its inner patio, with good shade in summer, its four walls covered with plants of all colors from the mountain of Libya. It had an air of ease and repose.

On an old wooden sign, beside which the Bedouins tied their camels, we read:

Inn of the Seven Sorrows.

"Seven sorrows," murmured Beremiz. "Strange! Do you by any chance know the owner of this inn?"

"I know him well," I replied. "He is a former rope merchant from Tripoli whose father served under Sultan Quervan. They call him the Tripolitan. He is very well thought of, for his simple, open nature, a fine and kindly man. They say that he went to the Sudan, in a caravan with soldiers of fortune, and brought back from African territory five black slaves who serve him with incredible fidelity. On his return, he gave up his rope business and opened his hostelry, with the help of the five slaves."

"With or without slaves," replied Beremiz, "this man, this Tripolitan, must be extremely original. He made the number 7 part of the inn's name, and 7 has always been, for Muhammadans, Christians, Jews, idolators, or pagans, a sacred number—the sum of 3, which is divine, and 4, which symbolizes the material world. That relation makes for strange connections among entities that total seven:

*Seven are the gates of hell.*
*Seven are the days of the week.*
*Seven wise men of Greece.*
*Seven the seas that cover the earth.*
*Seven the planets, and seven*
*the wonders of the world."*

He was continuing his strange and eloquent observations on the sacred number, when we saw at the door of the inn our good friend Sheik Salem Nasair, waving to us to approach.

"How happy I am to have found you at this moment, O counting man," said the sheik, smiling, as we drew near. "Your arrival is providential not just for me but also for these three friends who are here in the inn." He added, sympathetically but with keen interest, "Come in, come in! This is a most difficult matter."

He led us down a shadowy, damp corridor to the welcoming brightness of the inner patio, which had five or six round tables. At one of them sat three travelers.

When the sheik and the counting man approached, they raised their heads and made a salaam. One of them, who seemed very young, was tall, slender and clear-eyed and wore a bright yellow turban with a white band, in which sparkled a quite beautiful emerald. The two others were stocky and broadshouldered, with the dark skin of African Bedouins. Their clothes and appearance set them apart. They were deep in a discussion that, judging by their gestures, was perplexing them, as happens when a solution is difficult to find.

The sheik addressed the three of them: "Here is the esteemed master calculator." And, to Beremiz, he added, "Here are my three friends. They are sheep rearers from Damascus. They are facing one of the strangest problems I have come across. It is this: as payment for a small flock of sheep, they received, here in Baghdad, a quantity of excellent wine, in 21 identical casks:

*7 full*
*7 half full*
*7 empty*

They now want to divide these casks so that each receives the same number of casks and the same quantity of wine. Dividing up the casks is easy—each would receive 7. The difficulty, as I understand it, is in dividing the wine without opening them, leaving them just as they are. Now, calculator, is it possible to find a satisfactory answer to this problem?"

Beremiz, after pondering for two or three minutes, replied, "The division of the 21 casks, O Sheik, can be done without much complication. I am going to suggest the simplest possible solution. The first will receive

*3 full casks*
*1 half full*
*3 empty*

for a total of seven casks. The second will receive

*2 full casks*
*3 half full*
*2 empty*

for a total of 7 casks. The third will also receive 7 casks, in the same arrangement. According to my division, each party will acquire 7 casks and an equal quantity of wine. Let us say that a full cask of wine holds two portions, and a half-full cask one. According to the division, the first partner will receive

$$2+2+2+1$$

a total of seven units, and each of the others will receive

$$2+2+1+1+1$$

also adding up to 7. This proves that my suggested division is exact and just. Although the problem appears complicated, its numerical resolution presents no difficulty."

His solution was received with much enthusiasm not just by the sheik but also by the three men of Damascus.

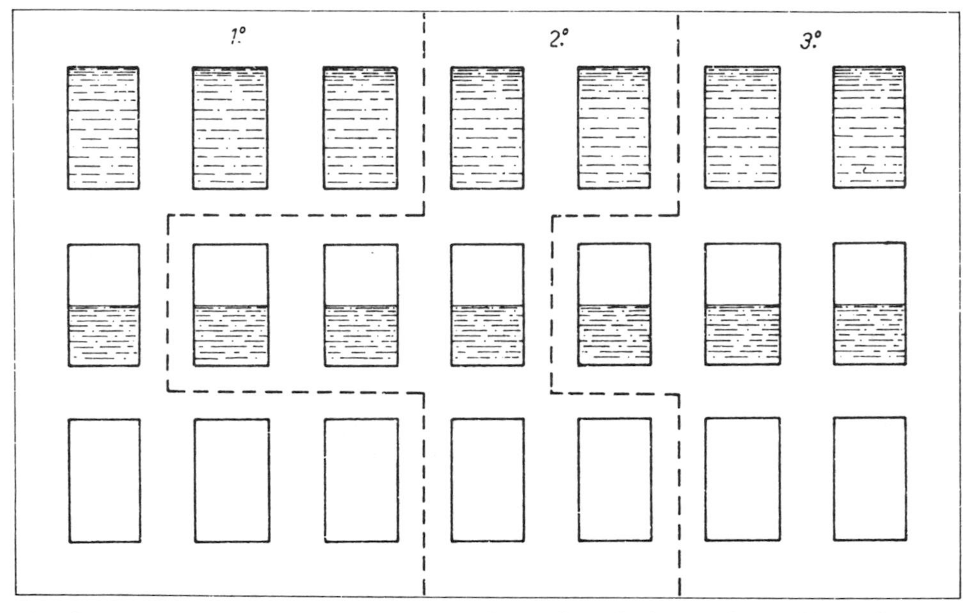


**This figure shows, in the simplest possible form, the solution to the problem of dividing the twenty-one casks of wine.**

"By Allah!" exclaimed the young man with the emerald. "This calculator is amazing! In a moment he cleared up a problem that seemed most difficult to us." Turning to the owner of the inn, he asked, in a friendly voice, "Man of Tripoli, what money have we spent at this table?"

"Your total bill, with your food, is 30 dinars," was the reply. Sheik Nasair wished to pay the bill, but the men of Damascus refused, which led to a small discussion and an exchange of compliments, with everyone speaking at once. At last it was agreed that Sheik Nasair, a guest, should pay nothing and that each of the others should pay 10 dinars; so 30 dinars were handed to a Sudanese slave for his master. A few moments later, the slave returned and said, "My master says he made an error. The bill is 25 dinars, and he has asked me to return 5 to you."

"That man of Tripoli is most honorable," remarked Sheik Nasair. And taking the five coins, he handed one to each of the three men, so that two remained. After exchanging a glance with the men from Damascus, the sheik handed them as a reward to the Sudanese slave who had served them food.

At that moment, the young man with the emerald rose and, looking gravely at his friends, said, "This business of paying over the 30 dinars has left us with a serious problem."

"Problem? I see no problem," replied the sheik, astonished.

"Oh yes," said the man from Damascus. "A serious and seemingly ridiculous problem. A dinar has disappeared. Think now. Each one of us paid 9 dinars. Three times nine is 27. Adding to these 27 the 2 that the sheik gave to the slave, we have 29 dinars. Of the 30 we handed over to the man from Tripoli, only 29 are accounted for. Where, then, is the other dinar? Where has it disappeared to?"

Sheik Nasair thought for a moment. "You are quite right, my friend. It seems clear. If each one of you paid 9 dinars, and the slave received 2, that adds up to 29. One of the original 30 is missing. How is that?"

At that moment, Beremiz, silent until then, intervened and addressed the sheik. "You are mistaken, O Sheik. The accounting should not be done in that way. Of the 30 dinars paid to the man from Tripoli for food, 25 went to the Tripolitan, 3 were returned, and 2 were a tip to the Sudanese slave. Nothing disappeared, and the account presents no problem. Of the 27 dinars paid over, the Tripolitan received 25, and the slave 2."

The men from Damascus, hearing Beremiz's explanation, exploded in loud laughter. "By the virtues of the Prophet," exclaimed the oldest of them, "this counting man has solved the mystery of the disappearing dinar and saved the reputation of this hostelry. Thanks be to Allah!"

# 9

# In the stars

*In which we receive a visit from Sheik Iezid, the poet. The strange results of an astrologer's predictions. Women and mathematics. Beremiz is invited to teach mathematics to a young girl. Her odd situation. Beremiz speaks of his friend and master, Nô-Elim the Wise.*

On the last day of the month Muharram, at sunset, the renowned Iezid Abul Hamid came looking for us at the inn.

"Some new problem to solve, O Sheik?" Beremiz asked with a smile.

"You guessed it, my friend!" replied our visitor. "I am faced with a serious difficulty. I have a daughter, Telassim, with a quick intelligence and an obvious leaning toward her studies. When Telassim was born, I consulted a famous astrologer who knew how to read the future by observing clouds and stars. He told me my daughter would be happy for her first eighteen years. From that age on, she would be threatened by a series of tragic misfortunes. He, however, had a way of keeping her bad luck from deeply affecting her destiny. Telassim, he said, ought to learn the properties of numbers and their many working possibilities. But to master numbers and calculation, it is essential to know the science of al-Khwarizmi, that is, mathematics. So I decided to provide a happy future for Telassim by making her study the mysteries of calculus and geometry."

The kindly sheik paused and then went on, "I looked for various scholars at court but could not find one able to teach geometry to a seventeen-year-old girl. One of them, an extremely gifted man, tried to dissuade me: 'Would you try to teach a giraffe to sing?' he asked. 'Its vocal chords cannot produce the least sound. It would be a dreadful waste of time, a useless labor. A giraffe will never sing. And the female brain,' said the holy man, 'is incapable of grasping the first principles of geometry. This unique science is founded in reason, in the use of equations, and in the application of clear principles with the help of logic and proportion. How can a girl who has been shut up in her father's harem learn algebraic formulas and geometric theorems? Never! Easier for a whale to make a pilgrimage to Mecca than for a woman to learn mathematics. Why take on the impossible? If bad luck must come our way, let it be by the will of Allah.' "

The sheik, looking serious, got up from his cushion and paced from side to side, before going on even more gloomily, "Discouragement, the great corrupter, settled on my spirit when I heard these words. Nevertheless, visiting my good friend the merchant Salem Nasair one day, I heard effusive reference to the new calculator from Persia who had arrived in Baghdad. I heard the story of the eight loaves, in all its detail, and it impressed me deeply. I set out to find the counting man and went specially to meet him at the house of the Vizier Maluf. I was astonished at how originally he solved the question of the 257 camels, reduced in the end to 256. Do you remember?"

And Sheik Iezid, raising his head and looking solemnly at the Man Who Counted, added, "Are you, my Arab brother, able to teach the ingenuities of calculating to my daughter Telassim? I will pay you any price you ask; and you can continue in your post as secretary to Vizier Maluf."

"Most generous Sheik," replied Beremiz at once. "I see no reason to refuse your noble invitation. In a few months I can teach your daughter the workings of algebra and the secrets of geometry. The philosophers are doubly in error in their estimation of the intellectual capabilities of women. The female intelligence, well directed, can perfectly encompass the beauties and secrets of science. It will be simple to disprove the unfair ideas of the holy men. History yields up various examples of women who distinguished themselves in mathematics. In Alexandria, for example, lived Hypatia, who taught hundreds of people the science of calculating, wrote a commentary on the work of Diophantus, analyzed the difficult texts of Apollonius, and corrected the astronomical tables currently in use. Be neither afraid nor uncertain, O Sheik. Your daughter will easily grasp the knowledge of Pythagoras, Allah be praised! It remains only to arrange the day and hour of her first lesson."

And the noble Iezid replied, "As soon as possible! Telassim is already seventeen, and I am anxious to free her from the woeful predictions of the astrologers." He added, "I must warn you, however, of a detail that is not unimportant. My daughter lives within the harem and has never set eyes on a man not of our family. She can attend her lessons in mathematics only from behind a thick curtain, with her face veiled, and with two family slaves in attendance. Given this condition, do you accept?"

"With relish," replied Beremiz. "Clearly the reserve and modesty of a young girl have value above algebraic formulas. The philosopher Plato had the following sign hung on the door of his school:

No One May Enter without Knowledge of Geometry

One day a depraved young man turned up, eager to attend Plato's Academy. The master, however, refused to admit him, declaring, 'Geometry is pure and simple. Your shamelessness is offensive to such a pristine science.' In this way, the famous pupil of Socrates demonstrated that mathematics cannot accompany depravity and is insulted in immorality. It will be a pleasure to instruct your daughter, whom I do not know, and whose face I will never have the good fortune to admire. If Allah is willing, I can begin tomorrow."

"Perfect," replied the sheik. "One of my servants will come for you a little after second prayers. Farewell."

When Sheik Iezid left the inn, I suggested to the Man Who Counted that the task might be beyond him. "One thing puzzles me, Beremiz. How can you teach mathematics to a young girl when you never really studied from books or attended the classes of the sages? How did you learn the power of calculation that you apply so brilliantly and opportunely? I know well that you uncovered its mysteries among sheep, fig trees, and flights of birds when you were a shepherd . . ."

"You are mistaken, my friend," Beremiz replied serenely. "While I was watching my master's flocks, in Persia, I knew an old dervish called Nô-Elim. I once saved his life during a violent sandstorm. From then on, he was my closest friend. He was a wise man, and he taught me useful and wonderful things. Having learned from such a man, I can teach geometry to the very last book of the unforgettable Euclid of Alexandria."

10

# Bird in the hand

*Our arrival at Iezid's palace. The ill-tempered Tara Tir questions Beremiz's calculations. Caged birds and perfect numbers. The Man Who Counted praises the sheik's kindness. We hear a tender and enchanting song.*

It was a little past four when we left the inn and made our way to the home of Iezid Abul Hamid. With the help of a pleasant and conscientious slave, we quickly maneuvered through the winding streets of the Muassan neighborhood and arrived at a magnificent palace set in the center of an elegant park.

Beremiz marveled at the distinguished appearance of Iezid's home. In the center of the park stood a great silvered tower on which rays of sunlight made rainbows of color. A large patio led into the main house through an iron gate etched in the most beautiful artistic detail. A second interior patio, in the center of which was a carefully ordered garden, divided the house into two wings. One of these contained the family's bedrooms; the other was given over to public rooms and the salon where the sheik often met with philosophers, poets, and viziers.

In spite of its decoration, the sheik's palace appeared sad and gloomy. Those who looked on the barred windows could not imagine the displays of art that lined the interior rooms. The two wings were joined by a long gallery supported by ten slender columns of white marble, punctuated with horseshoe arches and decorated with a plinth of tile in relief and a mosaic floor. Two splendid staircases, also of white marble, led to a garden with a large, peaceful pond surrounded by flowers of all colors and scents. A vast cage of birds, decorated with mosaics, seemed to be the centerpiece of the garden. There were birds of strange voice, rare breed, and brilliant plumage. Some, of unusual beauty, belonged to species unknown to me.

Our host greeted us very cordially as he joined us from the garden. There was within a dark, thin, broad-chested young man, whose manner revealed him to be lacking in courtesy. At his waist he sported a rich dagger with a hilt of ivory. He had a piercing, belligerent gaze, and his brusque, agitated way of speaking was most disagreeable.

"So, this is your calculator?" he snorted, disdain in his words. "What a trusting soul you are, Iezid. You are going to allow some passing beggar to approach and speak to the beautiful Telassim? That's the last thing you need! By Allah, how naive you are!" And he gave an ugly laugh.

His insolence infuriated me, and I wanted to answer this boor's rudeness with fisticuffs. Beremiz, however, remained cool. Or perhaps the Man Who Counted saw in this occasion, in these insults, only another problem to be solved.

The poet, disturbed by the man's inconsiderate behavior, said, "Mathematician, please forgive the rash judgment of my cousin Tara Tir. He does not know, and therefore cannot properly appreciate, your mathematical skills. He is, more than anything else, worried about Telassim's future."

"Of course, I don't know the mathematical skills of this stranger! I couldn't care less how many camels pass through Baghdad searching for shade and food," the young man exclaimed. And then, speaking rapidly, tripping over his words, he continued, "My cousin, in just a few minutes I can show that you are completely mistaken about the talents of this adventurer. If you will allow me, I will put an end to this man's science, using no more than a couple of trivialities I picked up from a master of Mosul."

"Yes, indeed. Go ahead," agreed Iezid. "You can ask the calculator questions right now, or raise any problems that you like."

"Problems? What for? Would you compare the science of the sages with that of a jackal in a cage?" he retorted crudely. "I assure you that it won't be necessary to come up with some problem to unmask this ignorant Sufi. I won't have to trouble my memory to achieve my result, much quicker than you think."

And, fixing Beremiz with a cold, unyielding gaze, he pointed to the large birdcage and asked, "Tell me, 'fowl calculator,' how many birds are in that cage?"

Beremiz Samir crossed his arms and studied the bright flock with intense concentration. It would be madness, I thought, to try to count the birds flying restlessly around the cage, dashing from one perch to another.

There was an expectant silence. After a few seconds, the Man Who Counted turned to the gracious Iezid and said, "I implore you, O Sheik, to free three of these birds immediately; then it will be both simpler and more pleasing to announce the total number."

His request seemed foolish. By logic, whoever could count a certain number could just as easily count three more. Most intrigued by such an unexpected request, Iezid ordered the bird keeper to do as Beremiz had suggested. Freed from captivity, three pretty hummingbirds arrowed into the sky.

"Now there are in this cage," said Beremiz in a deliberate manner, "496 birds."

"Wonderful," exclaimed Iezid enthusiastically. "The exact number! And Tara Tir knows it! I told him there were exactly 500 in my collection. I sent one nightingale to Mosul, and now we have freed 3, leaving 496."

"A lucky guess," grumbled Tara Tir, with a gesture of disgust.

Moved by curiosity, Iezid asked Beremiz, "Can you tell me why you prefer 496 when it would have been just as simple to add 496 and 3 to get 499?"

"I will explain, O Sheik," Beremiz replied proudly. "Mathematicians always try to give preference to remarkable numbers and avoid dull, ordinary sums. But between 499 and 496 there can be no doubt. The number 496 is a perfect number and therefore deserves our preference."

"And what do you mean by a perfect number?" the poet asked. "What makes a number perfect?"

"A perfect number" explained Beremiz, "is one that is equal to the sum of its divisors, excluding the number itself. So, for example, the number 28 has five divisors:

*1, 2, 4, 7, and 14*

The sum of the divisors

$$1+2+4+7+14$$

is exactly 28. Therefore 28 belongs in the category of perfect numbers.

"The number 6 is also perfect. The divisors of 6 are

*1, 2, and 3*

whose sum is 6.

"Along with 6 and 28, there is 496, which is also, as I said, a perfect number."

The ill-tempered Tara Tir, unable to bear Beremiz's further explanations, excused himself to Sheik Iezid and left muttering angrily, his rout by the skill of the mathematician not inconsiderable. As he passed in front of me, he shot me a glance of supreme scorn.

"I implore you, O calculator," the noble Iezid apologized again, "not to be offended by the words of my cousin Tara Tir. He is a hotheaded man, and since he took over the salt mines in al-Derid he has become irritable and violent. He has already suffered five murder attempts and several assaults from slaves."

It was evident that the intelligent Beremiz did not want to trouble the sheik, and he responded with kindness and generosity: "If we want to live in peace with our neighbors, we must restrain our anger and cultivate our goodwill. When I feel insulted, I try to follow the wisdom of Solomon: "A fool's wrath is presently known, but a prudent man covereth shame" (Proverbs 12:16). I can never forget the teachings of my good-hearted father. Whenever he saw me overwrought and ready for revenge, he would say, 'He who humbles himself before his fellow man is exalted in the eyes of God.' "

And after a short pause, he added, "I am, nevertheless, very grateful to the insolent Tara Tir, and I bear him no malice. His unruly nature gave me the opportunity to practice new acts of charity."

"New acts of charity?" the sheik replied with surprise. "What do you mean?"

"Every time we free a caged bird," explained Beremiz, "we accomplish three acts of charity. The first to the little bird, returning him to the freedom from which he has been snatched; the second to our own conscience, and the third to God . . ."

"You mean that if I were to free all of the birds in the cage . . ."

"I assure you, O Sheik, that you would be committing 1,488 acts of supreme charity!" exclaimed Beremiz, as if he already knew by heart the product of 496 and 3.

The generous Iezid was so impressed by the words of Beremiz that he decided to free all of the birds in the enormous cage. When they heard his orders, the servants and slaves were stunned. The collection, gathered with much patience and effort, was worth a fortune. It contained partridges, hummingbirds, brightly plumed pheasants, black gulls, ducks from Madagascar, Caucasian owls, and various swallows from China and India, exceedingly rare.

"Free the birds!" the sheik shouted again, waving his hand with its splendid ring.

The wide doors were opened, and the captive birds fled their prison in a flock, spreading over the treetops of the garden

Then Beremiz said, "Every bird, with its wings outspread, is a book, its pages open to heaven. An ugly crime, to rob or destroy this library of God."

And then we heard the first notes of a song. The voice was so soft and tender that it mingled with the trill of the little swallows and the soft cooing of the doves. At first the melody was enchanting and sad, filled with melancholy and longing, like the lament of a lonely nightingale. It rose then to a crescendo of complicated roulades, bright trills, and faltering cries of love that contrasted with the serenity of the afternoon and hung in the air like leaves lifted in the wind. Then it returned to the sad and mournful air of the beginning and seemed to hang over the garden like a whisper

*Though I speak with the tongues of men*
*and of angels,*
*and have not charity,*
*I am become as sounding brass,*
*or a tinkling cymbal,*
*I am nothing*
*I am nothing*

*And though I have the gift of prophecy,*
*And understand all mysteries, and all knowledge,*
*so that I could remove mountains,*
*and have not charity,*
*I am nothing*
*I am nothing*

*And though I bestow all my goods*
*to feed the poor,*
*and though I give my body to be burned,*

*and have not charity,*
*I am nothing*
*I am nothing*

The charm of that voice seemed to envelop the place in a wave of indefinable joy. Even the air seemed to lighten.

"It is Telassim singing," said the sheik, seeing that we were listening closely, captivated by that strange sound.

The birds were flying away, filling the air with their joyful songs of freedom. There were only 496, but they seemed like ten thousand.

Beremiz was wrapped in silence. The notes of the music penetrated his spirit, joining the happiness he felt at the liberation of the birds. Then he raised his eyes, searching for the place from where that voice came.

"And who wrote those beautiful verses?" he asked.

The sheik answered, "I don't know. A Christian slave taught them to Telassim, and now she cannot forget them. They must be from a Nazarene poet. That is what my wife told me, the mother of Telassim."

# 11

# For good measure

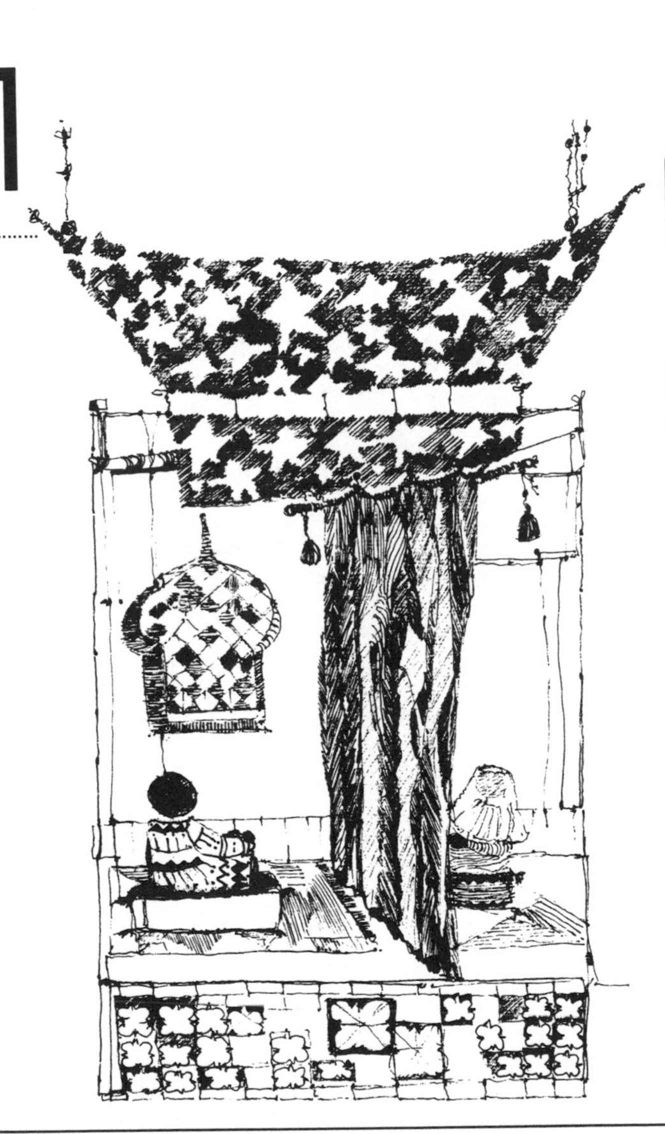

*How Beremiz started his lessons in mathematics. A phrase from Plato. The unity of God. What is measuring? The parts of mathematics. Arithmetic and numbers. Algebra and relations. Geometry and forms. Mechanics and astronomy. A dream of the King Assad Abu Carib. The "invisible student" prays to Allah.*

The room where Beremiz was to give his lessons was very spacious. It was divided by a thick, heavy curtain of red velvet that hung from the ceiling to the floor. The ceiling was brightly painted; the columns were golden. Over the carpets were scattered large silk cushions, the edges of which were inscribed with verses from the Koran. The walls were decorated with fanciful blue designs intertwined with beautiful lyrics by Antar, the desert poet. In the center of the room, between two columns, in gold letters against a blue background, were these words from a paean by Antar:

*When Allah loves one of his followers,*
*He leads him to inspiration.*

A gentle perfume of incense and roses pervaded the room as dusk approached. The polished marble windows were open and looked out toward the garden and rows of luxuriant apple trees stretching down to the gray, rough waters of a river. A black slave stood by the door, her face exposed, her fingernails colored with henna.

"Is your daughter here?" Beremiz asked the sheik.

"Of course," replied Iezid. "I told her to sit on the other side of the room, behind the curtain, where she will be able to see and hear. Nevertheless, she will be invisible to everyone on this side."

Truly, everything had been arranged so that one could not make out even the silhouette of the young girl who was to be Beremiz's pupil. Perhaps she watched us through some tiny hole in the velvet, imperceptible to us.

"I think that now we can begin," the sheik said, continuing affectionately, "try to pay attention Telassim, my daughter."

"Yes, Father," answered a well-bred female voice on the other side of the room.

Then Beremiz got ready to begin his lesson: he crossed his legs and sat down on one of the cushions in the middle of the room. I discreetly found a place in the corner and made myself as comfortable as I could. The sheik sat by my side.

All science proceeds from prayer. Therefore, Beremiz began his class with a prayer: "In the name of Allah, the All-Merciful. Praised be the omnipotent Creator of the universe! The mercy of God is our supreme gift! We adore you, O Lord, and implore your grace! Lead us in the path of righteousness, in the path of those blessed by your hand!"

When he had finished the prayer, Beremiz said, "When we look to the sky on a still, clear night, we feel that we are incapable of understanding the wondrous works of God. Before our astonished eyes, the stars form a luminous caravan that files across an infinite desert, circling the immense nebulas and planets, following eternal laws, from the depths of space, and suggesting to us a very precise notion: the idea of 'number.'

"There lived in Greece, when that country was still pagan, a philosopher named Pythagoras—How wise is Allah! When a disciple asked him which were the dominant forces in the affairs of man, the wise man answered, 'Numbers rule everything.'

"Truly. The simplest thought cannot be achieved without including, in many aspects, the fundamental concept of number. The Bedouin in the desert, his head bowed in prayer, murmurs the name of God, and his spirit is filled with a number: 'Oneness, Unity.' Yes, God, according to the truth written in the pages of the Holy Book and repeated by the Prophet, is one, eternal, and immutable! Therefore, the number appears, in the framework of our intelligence, as a symbol of the Creator.

"From numbers, which are the basis of all reason and understanding, comes another notion of indisputable importance: the notion of 'measurement.'

"To measure is to compare. Nevertheless, only those numbers that contain an element on which to base comparison are susceptible to measurement. Would it be possible to measure the vastness of space? Not at all. Space is infinite and therefore has no basis for comparison. Is it possible to measure eternity? Never. In human terms, time is always infinite, and in the calculation of eternity nothing so ephemeral can serve as a unity of measurement.

"In many cases, nevertheless, it would be possible for us to substitute one dimension that does not fit with our system of measurements for one that can be calculated with more certainty. This exchange, in order to simplify the measuring process, constitutes the principal object of the science that we call *mathematics.*

"In order to reach his goal, the mathematician has to study numbers, their properties and permutations. This aspect is called *arithmetic.* Once numbers are understood, it is possible to use them to evaluate dimensions that vary or that are not known but that can be symbolized with formulas and equations. In that way, we get *algebra.* The measurements we make of reality are represented by material things or symbols; in either case, these things or symbols are endowed with three attributes: form, size, and position. It is important, therefore, to study such attributes. That is the objective of *geometry.*

"Mathematics also deals with the laws that govern movement and forces, laws that appear in the admirable science called *mechanics.* Mathematics puts all of its wonderful resources at the disposal of a science that raises the spirit and exalts men. That science is *astronomy.*

"Some think that, within the framework of mathematics, arithmetic, algebra, and geometry are entirely distinct disciplines: that is a grave error. They all work together, the one helping the other, in some cases even interchangeable.

"Mathematics, which teaches man to be simple and humble, is the basis of all the arts and sciences.

"Let me repeat to you an incident that occurred to a famous Yemenite monarch: Assad Abu Carib, king of Yemen, while resting one day in the spacious balcony of his palace, dreamed that he had met seven young maids walking along a path. At a certain point, overcome with fatigue and thirst, they stopped under the burning desert sun. Just then there rose up a beautiful princess who offered the pilgrims a pitcher of water. The kind princess quenched their thirst, and they continued on their journey reinvigorated.

"When he awoke, Assad Abu Carib was so impressed by this inexplicable dream that he decided to call for a famous astrologer, named Sanib, to consult him on the meaning of this vision that he—a just and powerful king—had seen in the world of image and fantasy. The astrologer Sanib said, 'Sir, the seven young maids walking along the path were the divine arts and the human sciences: Painting, Music, Sculpture, Architecture, Rhetoric, Dialectics, and Philosophy. The kind princess who came to their aid was the great and prodigious science of Mathematics.' 'Without the help of mathematics,' the wise man continued, 'the arts could not advance and all the sciences would perish.' Moved by these words, the king decided to organize centers for the study of mathematics in all the cities, villages, and oases of his country. By the sovereign's order, able and eloquent wise men went forth into the bazaars, inns, and caravansaries to give lessons in arithmetic to the traders and nomads. Within a few months the country had become more prosperous. In parallel with the advance of science, the real wealth of the country was growing; the schools were filled with students; commerce expanded rapidly; works of art increased; monuments were raised; rare and foreign treasures abounded in the cities. The country of Yemen opened itself to progress and wealth, but then misfortune!—*Maktub*! This prodigious flowering of work and wealth came to an end. King Assad Abu Carib closed his eyes to the world and was raised by the infidel Asrail to the heaven of Allah. The death of the monarch opened two graves: one received the body of the glorious king, and in the other was buried the artistic and scientific culture of his people. A vain, indolent prince, a man of little intellectual merit, ascended to the throne. He spent more time on idle pursuits than on the administrative problems of his country. In just a few months, all of the public services had become chaotic, schools were closed, and artists and wise men were forced to flee from the threats of malefactors and thieves. The public treasury was criminally squandered on idle festivals and lavish banquets. The country was brought to ruin by mismanagement and was finally attacked by eager enemies, who quickly conquered the Yemenites.

"The story of Assad Abu Carib shows that the progress of a people is linked to the development of their mathematical abilities. In all the universe, mathematics is number and measure. Oneness, the symbol of the Creator, is the beginning of everything, which would not exist but for the unvarying proportions and relations of numbers. All of life's great enigmas can be reduced to simple combinations of either variable or constant, known or unknown, elements that we can solve.

"So that we can understand this science, we must begin with numbers. We will see how to examine them, with the help of Allah, the All-Merciful!

*"Uassalan!"*

With these words the Man Who Counted ended his first class. And then, as an agreeable surprise, we heard the voice of the invisible student, hidden behind the velvet curtain, speaking the following prayer:

"O omnipotent God, Creator of heaven and earth, forgive the poverty, the meanness, the naïveté of our hearts. Listen not to our voices but to our inarticulate cries; attend not to our desires but to the clamor of our needs. How many times do we ask for something that could never be ours!

“God is great!

“O God! We thank you for this world, our great home, its size and wealth, the multifarious life of the world of which we are a part. We praise you for the splendor of the blue skies and for the evening breeze and for the clouds and for the stars in the heavens. We praise you, Lord, for the immense oceans, for the water that runs in the streams, for the eternal hills, for the luxuriant trees, and for the carpet of grass that soothes our feet.

“God is merciful!

“We thank you, Lord, for the many delights that make us feel in our spirit the beauty of life and love . . .

“O God, the All-Merciful! Forgive the poverty, the meanness, the naïveté of our hearts.”

# 12

# Circular reasoning

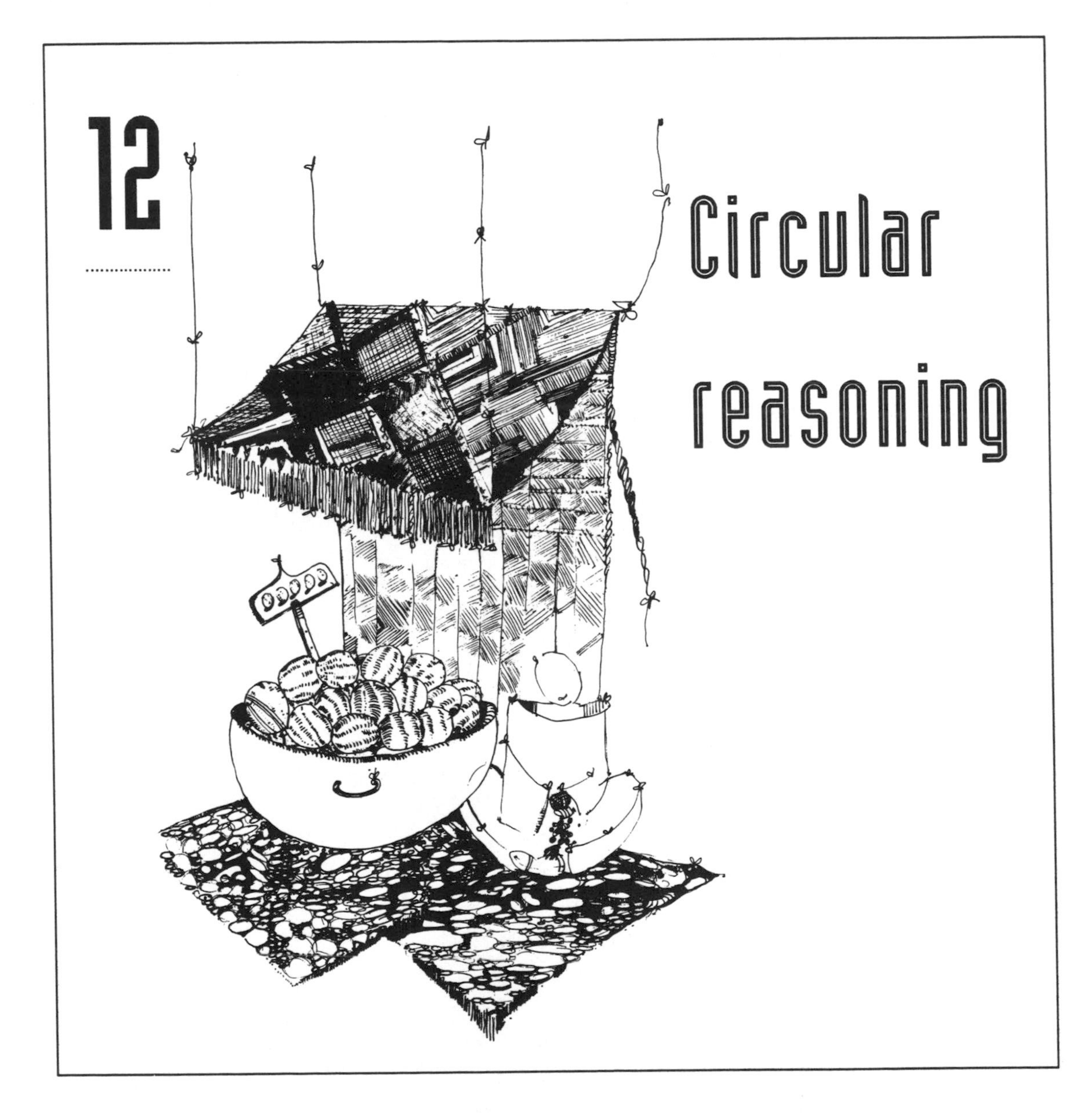

*Beremiz reveals his fascination with skip rope. The curve of the* marazan *and the spiders. Pythagoras and the circle. Our meeting with Harim Namir. The problem of the sixty melons. How the magistrate lost a bet. The voice of the blind muezzin calls believers to prayers at sunset.*

It was close to the time of prayer when we left the poet Iezid's sumptuous palace. Passing the shrine of Ramih, we heard birds twittering in the branches of an old fig tree. "Look! These are certainly some of the birds set free today," I said to Beremiz. "How pleasant to hear their joy at recovering their freedom being turned into song!" At that moment, however, Beremiz was not interested in the singing of the birds. He was absorbed in watching some children playing in a nearby street. Two of them held the ends of a slender rope that was about four or five cubits long. The others were trying to jump over it while the two turned it lower or higher, depending on the skill of the jumper.

"Look at the rope, my Baghdad friend!" said the calculator, taking my arm. "Look at its perfect curve. Do you not think it worthy of study?"

"What are you talking about? The rope?" I exclaimed. "I don't see anything extraordinary in this game of children taking advantage of the last light of day for their enjoyment."

"Well then, my friend, your eyes are blind to the beauty and wonder of nature. When the children raise the rope, holding it by its ends and then letting it fall freely from its own weight, the rope forms a curve on its own, as a result of natural forces. On other occasions I have observed this same curve, which the sage Nô-Elim called *marazan,* in cloth and in the humps of some dromedaries. Is this curve analogous to the parabola? In the future, if Allah wishes, geometers will find ways to trace this curve, point by point, and will rigorously study its properties."

"Nevertheless," he continued, "there are many other, more important curves. First the circle. Pythagoras, the Greek philosopher and mathematician, considered the circle the most perfect curve, thereby associating the circle with perfection. And the circle, being the most perfect curve of all, is the easiest to draw."

At that point Beremiz interrupted the lecture he had just started on curves, indicated a young man a short distance away, and shouted, "Harim Namir!"

The young man turned around abruptly and came toward us, smiling. Then I saw that he was one of the three brothers we had met in the desert arguing over the inheritance of thirty-five camels—a difficult division, complicated with thirds and ninths, that Beremiz solved by resorting to a curious trick that I have already explained.

"Destiny has led us to the great calculator. My brother Hamed is trying to clear up a matter of sixty melons that no one can resolve." And Harim led us to a house where we found his brother Hamed Namir with several merchants.

Hamed was very pleased to see Beremiz and, turning to the merchants, said, "This man who has just arrived is a great mathematician. Thanks to his valuable help, we found a solution to a problem that seemed insoluble: how to divide thirty-five camels among three people. I am certain that he will be able to explain within minutes the discrepancy we have encountered in the sale of sixty melons."

Beremiz was carefully apprised of the case. One of the merchants explained, "The two brothers Harim and Hamed brought me two batches of melons to sell in the market. Harim brought me 30 melons that were to be sold at a price of 3 for 1 dinar; Hamed also brought me 30 melons, but they were to be sold at the higher price of 2 melons for 1 dinar. Logically, once the melons were sold, Harim would receive 10 dinars and his brother 15. The total sale price would then be 25 dinars.

"Yet, when I reached the marketplace, doubts began to trouble my spirit. If I started by selling the more expensive melons, I would lose my buyers. If I started by selling the cheaper melons, I thought, I would later have trouble in selling the other thirty. The only solution was to sell the two batches at the same time.

"Having made this decision, I put all the melons together and started to sell them in lots of 5 for 2 dinars. The reasoning seemed to me clear: if I sold 3 for 1 dinar and then 2 for 1, it would be easier to sell 5 for 2 dinars.

"After selling 60 melons in 12 lots of 5 each, I made 24 dinars. But how could I pay both brothers if one was to get 10 and the other 15 dinars? There was a difference of 1 dinar. I don't know how to explain this difference since, as I said, the business was handled with the utmost care. Isn't it the same thing to sell 3 for 1 and then 2 for 1 as to sell 5 for 2?"

"The matter would have not been at all important," broke in Hamed Namir, "if it had not been for the absurd meddling of the magistrate who supervises the marketplace. Having heard of this matter, the magistrate did not know how to account for the difference, and he bet 5 dinars that it was caused by the theft of one melon during the sale."

"The magistrate is wrong," said Beremiz, "and he will have to pay the 5 dinars. This is how to explain the difference:

"Harim's batch consisted of 10 lots of 3 melons each. Each lot should have been sold for 1 dinar. The total would have been 10 dinars.

"Hamed's batch consisted of 15 lots of 2 melons each, which, at a price of 1 dinar per lot, would have brought a total of 15 dinars.

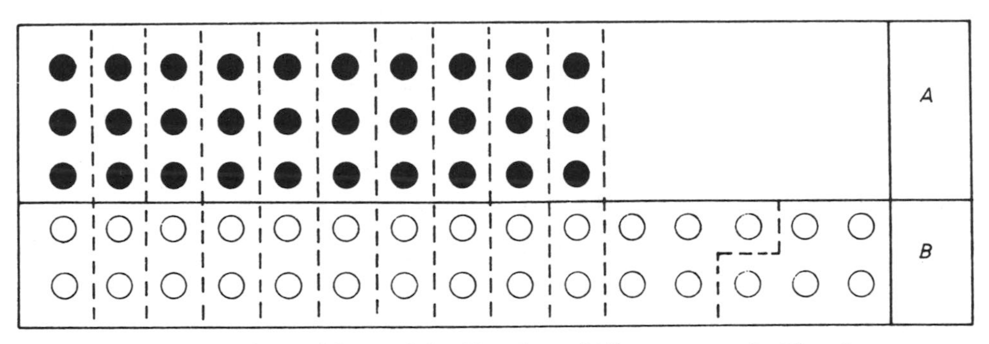


**This figure clears up the problems of the 60 melons. "A" represents the 30 melons which were sold at 3 for a dinar and "B" the 30 melons at 2 for a dinar; the whole shows them divided into twelve batches of 5 melons each, sold for 2 dinars.**

"Notice that the number of lots in the two batches is not the same. In order to sell the melons in lots of 5, only 10 lots of 5 could have been sold for 2 dinars; these 10 lots having been sold, there would have remained 10 melons belonging only to Hamed's batch and these, being more expensive, would have had to be sold at a rate of 2 for 1 dinar.

"The difference of one dinar was, then, a result of the sale of the last 10 melons. Therefore, there was no theft. The loss of 1 dinar was a result of the difference in price between the two batches."

At that moment we had to break off our meeting. The voice of the muezzin, echoing through the air, was calling the faithful to afternoon prayers.

*"Hai al el-salah! Hai al el-salah!"*

Without losing a moment, each of us began our ritual preparations, in accordance with the Holy Book. The sun had just reached the horizon. It was the hour of *maghrib,* the prayer at dusk. From the tower of the Mosque of Omar, the blind muezzin called the faithful to prayer in a deep, deliberate voice:

"Allah is great and Muhammad, the Prophet, is the true messenger of God. Moslems, come to pray! Come to pray! Remember that everything is dust, except Allah!"

The merchants, following Beremiz, unrolled their bright rugs, removed their sandals, turned toward the Holy City, and cried out:

"Allah, wise and merciful! Praised be the omnipotent Creator of heaven and earth! Lead us in the path of righteousness, in the path of those who are by you protected and blessed!"

# 13

# Friendship knows no bounds

*Which tells of our visit to the caliph's palace, and of the audience he was kind enough to grant us. Of poets and friendship—friendship between men and between numbers. The Man Who Counted is extolled by the caliph of Baghdad.*

Four days later, in the morning, we were informed that we would be received in solemn audience by Caliph Abul Abas Ahmed al-Mutasim Billah, emir of all believers, vicar of Allah. That message, so gratifying to any Muslim, had been eagerly awaited not only by me but by Beremiz as well.

It is possible that the sovereign, hearing from Sheik Iezid of some of the feats performed by the renowned mathematician, might have shown interest in getting to know the Man Who Counted. Nothing else can explain our presence in the court among the most distinguished figures in Baghdad society.

Entering the rich palace of the emir, I was dazzled. The great arcades, curving harmoniously and supported by tall and slender pairs of columns, were decorated around their bases with fine mosaic. I could see that these mosaics were made up of tiny red and white tiles, set in stucco. The ceilings of the principal rooms were decorated in blue and gold; the walls of all the rooms were covered with embossed tiles and the walks with mosaic. The lattices, the carpets, the divans—in fact, all the furnishings of the palace—showed off the supreme magnificence of a prince of Hindu legend.

Outside, in the gardens, the same magnificence was visible, enhanced by the hand of nature, perfumed by a thousand different aromas, covered in a carpet of greenness, bathed by the river, refreshed by innumerable white marble fountains, beside which thousands of slaves labored.

When we arrived, we were led to a divan by one of the aides of Vizier Ibrahim Maluf. We saw the all-powerful monarch seated on a rich throne of ivory and velvet. The overpowering beauty of the great room disturbed me somewhat. All the walls were decorated with fine inscriptions, put there by the cunning hand of an inspired calligrapher. I noted that those inscriptions were verses from our most brilliant poets. Everywhere stood urns filled with flowers—on the cushions, flowers without petals, flowers woven in the carpets or exquisitely worked into the gold platters.

Many sumptuous columns caught my eye, their caps and shafts elegantly worked by the chisels of Moorish artists, unmatched masters of the many geometric variations in the design of flower and leaf, of tulips, of lilies, of a multitude of diverse plants, wonderfully harmonious, indescribably beautiful.

In attendance were seven viziers, two judges, several sages, and various well-known dignitaries.

It fell to the noble Maluf to present us. The vizier, elbows at waist, his slender hands extended, palms open, spoke as follows: "To comply with your command, O King of the Temple, I made sure that today, in this illustrious audience, my present secretary, Beremiz Samir, and his friend Hanak Tade Maia, scribe and functionary of the palace, should present themselves."

"Then be welcome, Muslim brothers!" replied the sultan in a warm and friendly voice. "I admire men of wisdom. Anyone versed in mathematics, under the skies of this land, can always count on my support and, indeed, my firm protection."

"May Allah guide you, O Master!" exclaimed Beremiz, bowing.

I remained still, my head bowed, my arms crossed, but since I had not been addressed, I did not presume to answer.

The man who held in his hands the destiny of all Arabs seemed both open and generous. His features were fine, tanned by the desert sun, prematurely furrowed. When he smiled, which he did often, he showed regular white teeth. He wore at his waist, in a silk girdle, an elegant dagger, its sheath studded with precious stones. His turban was green, with thin white stripes. Green, as everyone knows, marks the descendants of the Holy prophet Muhammad, to whom all glory and peace!

"I have many important things to discuss in this audience," declared the caliph. "I have no wish, however, to embark on a discussion of serious matters of politics without first having some clear proof that the Persian mathematician commended by my friend the poet Iezid is truly a worthy and accomplished calculator."

So appealed to by the illustrious monarch, Beremiz felt compelled to justify Sheik Iezid's confidence in him. Addressing the sultan, he spoke, "O ruler of the faithful, I am no more than a simple shepherd who has been graced by your regard."

After a short pause, he went on, "Even so, my generous friends consider it justified to name me among the men of numbers, and I feel flattered by such high praise. I think, however, that men in general are good calculators. So is the soldier on campaign who estimates distance in a single glance. So is the poet who counts syllables and verifies the cadence of his verses. So is the musician who applies in his compositions the laws of perfect harmony. So is the painter who draws with the unvarying proportions of perspective in mind. So is the humble rug maker who arranges one by one the threads of his labor. All of these, O King, are fine and accomplished calculators."

After turning his noble gaze on those who surrounded the throne, Beremiz continued, "I see to my infinite pleasure that you, Master, have wise and learned men all around you. I see in the shadow of your mighty throne distinguished men who pursue their studies and extend the frontiers of science. The company of wise men, O King, is for me the greatest of treasures. The worth of a man lies in what he knows. Knowledge is power. Wise men teach by example, and there is nothing that captures the human spirit more convincingly than example. Man, however, should not pursue knowledge except to do good.

"The Greek philosopher Socrates lent all the weight of his authority to make this point: 'The only useful knowledge is that which betters us.' Seneca, another celebrated thinker, asked in disbelief, 'What does it matter to know what a straight line is if one has no notion of rectitude?' Allow me then, O generous and just King, to pay homage to the wise and learned ones in this room."

The Man Who Counted paused briefly and went on, solemn and eloquent, "In my daily labors, taking notice of all that Allah brought from nonbeing into being, I learned to value numbers and to use them in accordance with practical and reliable laws. I feel, however, some difficulty in providing the proof you have just asked for. Still, counting on your celebrated generosity, I must nevertheless say that in this splendid divan I see nothing but admirable and eloquent demonstrations of mathematics everywhere. The walls of this elegant room are decorated with various poems, each one of precisely 504 words. Some of these words are written in black characters, the rest in red. The calligrapher who traced out the letters of these poems, disposing the 504 words, has shown that he has a talent and imagination equal to that of the poets who wrote these immortal verses."

"Indeed, O King," Beremiz went on. "And the reason is simple. I find, in these incomparable verses that grace this magnificent room, great praise of friendship. There, close to the column, I can read the first line of the famous poem of al-Muhalhil:

> If my friends depart me, most wretched will I be, since all my treasures depart me.

And there, farther on, I read the line of Tarafa:

> The enchantments of life lie solely in the great friendships we form.

Indeed, all these are sublime, profound, and eloquent. The greater beauty, however, lies in the ingenuities of the calligrapher, in demonstrating that the friendship these verses extol exists not just among those endowed with life and feeling. Friendship occurs also among numbers.

"You will ask, no doubt, how to recognize among numbers those bound together in mathematical friendship. How can geometry distinguish in a numerical series those that are so bound?

"I will explain as briefly as I can the idea of friendship among numbers. Take, for example, the numbers 220 and 284; 220 is exactly divisible by the following numbers:

*1, 2, 4, 5, 10, 11, 20, 22, 44, 55, and 110*

In turn, 284 is exactly divisible by these numbers:

*1, 2, 4, 71, and 142*

"Among these two numbers, there are remarkable coincidences. If we add up the numbers above that go into 220, we get 284; and if we add those that go into 284, we get exactly 220.

"From this result, mathematicians have concluded that the numbers 220 and 284 are 'friends,' that each of them seems to serve, delight, guard, and honor the other."

And he finished thus: "Now, O generous and just King, the 504 words that make up the poems in praise of friendship were traced in the following way: 220 of them in black characters and 284 in red characters, numbers that, as I explained, are friends.

"Now, notice another, no less interesting connection. The 504 words are formed in 32 lines, as you see. Well, the difference between 284 and 220 is 64, a number that, apart from being a square and a cube, is exactly twice the number of lines.

"A skeptic would call it mere coincidence, but a believer who follows the teachings of the Holy Prophet Muhammad—to whom all prayer and peace!—knows that so-called coincidences are impossible, unless Allah has already inscribed them in the Book of Destiny. I declare, then, that the calligrapher, dividing the number of 504 in two—220 and 284—wrote a poem on friendship that must move all men of spirit."

Hearing the words of the Man Who Counted, the caliph was overjoyed. It seemed incredible to him that he had counted at a glance the 504 words of thirty poems and verified that 220 were black and 284 were red.

"Your words, O man of numbers, have proved to me beyond doubt that you are in truth a mathematician of the highest worth. I have been delighted by that absorbing connection you call numerical friendship, and I am now interested in finding which calligrapher it was who inscribed the verses that adorn the walls of this room. It is quite easy to find out whether the disposition of the 504 words into two friendly parts was deliberate or a trick of destiny, the work of Allah, the Exalted One."

And, summoning one of his secretaries, Sultan al-Mutasim asked him, "Do you remember, Nuredin Zarur, which calligrapher worked in this palace?"

"I know him well, Master. He lives close to the Mosque of Ottoman," replied the sheik.

"Bring him here as soon as possible," the caliph ordered. "I want to question him at once."

"I hear and obey!"

And the secretary left like an arrow to carry out the order of the sovereign.

# 14

# An eternal truth

*Of the things that occurred in the throne room. The musicians and twin dancers. How Beremiz could distinguish Iclimia from Tabessa. An envious vizier criticizes Beremiz. The Man Who Counted praises theorists and dreamers. The king proclaims theory the victor over the demands of the immediate present.*

After Sheik Nuredin Zarur, the emissary of the king, went in search of the calligrapher who had written out the poems that decorated the room, there entered five Egyptian musicians who performed with great feeling the most tender Arab songs and melodies. While the musicians sang and played their harps, zithers, and flutes, two gracious dancers, by their appearance Spanish slaves, danced on a wide, circular stage for the pleasure of the assembled company.

Slave girls who are to become dancers are chosen carefully and are particularly valued, for they provide beauty and entertainment, personal satisfaction, and flattery for the guests. The dances vary according to the origin of the dancers, and their variety indicated the wealth and power of our host. The physical likeness of the two dancers was considered a high virtue. In order to find such a pair, a careful and exacting selection was required.

The similarity of the two slave girls astonished everyone. Both had the same slender waist, the same dark complexion, eyes outlined identically in black kohl; both wore the same necklaces, bracelets, and collars, and, to compound the confusion, there was not the slightest difference in their gowns.

At a certain moment, the caliph, who appeared to be in fine humor, addressed himself to Beremiz.

"What do you think of my pretty slaves? You will already have noticed that they are identical. One is named Iclimia and the other Tabessa. They are twins and are worth a fortune. I have never found anyone who could distinguish the two when they appear on stage. Look closely. Iclimia is on the right now. Tabessa is on the left, next to the column, giving us her best smile. By the color of her skin and the delicate perfume she exudes, she seems to be a leaf of aloe."

"I confess, O Sheik of Islam," Beremiz replied, "that these dancers are truly marvelous. Praised be Allah, the One God, who created beauty from which comes such seductive feminine creatures. Of beautiful women, the poet says:

> It is for your luxury that poets weave a fabric of gold thread and for your beauty painters create fresh immortality.
>
> To adorn you, to clothe you, to make you more exquisite, the sea gives its pearls, the earth its gold, the garden its flowers.
>
> Over your youth, the desire of man's heart spreads its glory."

"It seems to me, however, easy enough," offered the Man Who Counted, "to distinguish Iclimia from her sister Tabessa. One need only note their dresses."

"How is that possible?" replied the Sultan. "There is not the slightest difference between their dresses. Both, by my orders, wear identical veils, blouses, and dance skirts."

"Please forgive me, O generous King," Beremiz replied politely, "but the seamstresses did not respect your orders with the proper caution. Iclimia's skirt has 312 fringes while Tabessa's has 309. This difference in the total number of fringes is enough to dispel any confusion between the twin sisters."

On hearing this, the sultan clapped several times, stopped the dancing, and ordered the *hakim* to count the fringes on the dancers' costumes one by one.

Beremiz's calculations were confirmed. The lovely Iclimia had 312 fringes on her costume and her sister Tabessa only 309.

"By Allah!" exclaimed the caliph. "Even though he's a poet, Sheik Iezid did not exaggerate. This Beremiz is truly a talented counting man. He counted all the fringes on both costumes while the dancers twirled dizzily across the stage. By Allah! It's incredible!"

But when envy takes hold of a man, it opens his soul to every despicable, crude inclination.

There was in the court of al-Mutasim a vizier named Nahum ibn-Nahum, an envious, bad man. As he watched Beremiz's prestige rise before the caliph like a wave of sand propelled by a desert wind, the sheik was driven to despair and decided to catch out my friend and make a fool of him. So he approached the king and slowly spoke thus:

"I have just noticed, O Emir of the Creators, that the Persian calculator, our guest of this afternoon, is a gifted counter of elements and serial figures. He counted the 500 or so words written on the walls, indicated the friendship between numbers, spoke of their difference—64, which is both a cube and a square—and ended by counting one by one the fringes on the gowns of the beautiful dancers.

"It would be terrible if our mathematicians spent their time on such childish things without any practical use whatsoever. Really, what good is it to know that in the verses that we adore there are 220 plus 284 words? The concern of those who admire a poet does not lie in counting the letters of the verses or in counting the number of black or red words in his poems. Neither does it matter whether we know that in the gown of this beautiful and graceful dancer there are 312, 309, or even 1,000 fringes. This is all ridiculous and of very limited interest for those men of feeling who cultivate beauty and art.

"The clever man, supported by science, must dedicate himself to the solution of life's great problems. Wise men, inspired by Allah, the Exalted One, did not erect the dazzling edifice of mathematics so that this noble science would be used in the manner of this Persian counting man. It seems to me a crime to reduce the science of Euclid, Archimedes, or the extraordinary Omar Khayyám, whom Allah secures in his glory, to the pitiful task of merely enumerating things and beings. We are curious to see whether this Persian calculator can apply the talents he is said to possess to the solution of genuine problems, that is, problems that bear on the demands of everyday life."

"I believe you are slightly mistaken, Vizier," Beremiz replied quickly, "and I would be honored if you would allow me to clarify this insignificant error, and therefore I beg the generous caliph, our soul and master, to allow me permission to continue speaking."

"It seems to me there is a certain wisdom in Vizier Nahum ibn-Nahum's criticism," replied the caliph. "I believe that a clarification of the matter is absolutely required. So speak. What you say will shape the opinions of those who listen to you here."

There was a long silence in the room. Then the Man Who Counted spoke: "Learned men, O King of the Arabs, know that mathematics arose from the awakening of the human soul. But it was not born with utilitarian purposes. The first impulse of this science was the desire to solve the mystery of the universe. Its development came, therefore, from the effort to penetrate and understand the infinite. And even now, after centuries of trying to part the heavy veil, it is the search for the infinite that moves us forward. The material progress of man depends on abstract investigations and on present-day scientists, and the material progress of humanity in the future will depend on these men of science who work toward purely scientific ends, without considering the practical application of their theories."

Beremiz paused briefly and then continued, with a smile, "When the mathematician makes his calculations or looks for new relations among numbers, he does not look for truth with a practical purpose. To cultivate science only for its practical purpose is to despoil the soul of science. The theory that we study today, and that appears to us impractical, might have implications in the future that are unimaginable to us. Who can imagine the repercussions of an enigma through the centuries? Who can solve the unknowns of the future with the equations of the present? Only Allah knows the truth. And it is possible that the theoretical investigations of today may provide, within one or two thousand years, precious practical uses.

"It is important to bear in mind that mathematics, besides solving problems, calculating areas, and measuring volumes, also possesses much more elevated purposes. Because it is so valuable in the development of intelligence and reason, mathematics is one of the surest ways for a man to feel the power of thought and the magic of the spirit.

"Mathematics is, in conclusion, one of the eternal truths and, as such, raises the spirit to the same level on which we contemplate the great spectacles of nature and on which we feel the presence of God, eternal and omnipotent. As I have said, O illustrious Vizier Nahum ibn-Nahum, you have made a slight error. I count the verses of a poem, calculate the height of a star, measure the size of a country or the force of a torrent, and thus I apply the formulas of algebra and the principles of geometry, without concerning myself with the profit I might earn from my calculations and studies. Without dreams or imagination, science is impoverished. It is lifeless."

The nobles and wise men around the throne were deeply affected by Beremiz's eloquent words. The king approached the Man Who Counted, raised his right hand, and, with great authority, exclaimed, "The belief of the scientist-dreamer has triumphed and will always triumph over the vulgar opportunism of the ambitious scientist without philosophical belief. O word of Allah!"

On hearing these words, so justly and righteously spoken by the king, the hateful Nahum ibn-Nahum bowed, saluted the king, and, without saying a word, left the salon with his head bowed.

Correct was the poet who wrote:

*Let the imagination soar*
*Without illusion, what would life be?*

# 15 Squared away

*Nuredin, the messenger, returns to the caliph's palace. What he learned from a holy man. How the poor calligrapher lived. The number square and the chessboard. Beremiz discourses on magic squares. The question of the wise man. The caliph asks Beremiz to tell the story of chess.*

Nuredin had no luck in his quest. The calligrapher whom the king wished to question on the matter of friendly numbers was nowhere to be found in Baghdad. Telling of the steps he had taken to carry out the caliph's orders, Nuredin spoke as follows:

"I left the palace with three guards, heading for the Mosque of Ottoman—may Allah be praised once more! An old holy man who takes care of the mosque told me that the man I was looking for had lived for several months in a house nearby. A few days earlier, however, he had left for Basra with a caravan of rug merchants. He also told me that the calligrapher, whose name he did not know, lived alone and ventured out very rarely from his small and modest dwelling. I thought it wise to visit his old house in case I might find some clue as to where he had gone.

"The house had been abandoned since its previous inhabitant departed. Everything indicated the sorriest poverty. A broken bed in a corner was its only furnishing. On a rough wooden table, however, there was a chessboard with some pieces and, on the wall, a square filled with numbers. I found it strange that a man so impoverished, living such a meager life, would be a chess player and would have on his wall as decoration mathematical symbols. I decided to bring the chessboard and the number square with me so that our worthy sages might study these clues left behind by the old calligrapher."

The sultan, his interest awakened, ordered Beremiz to take a close look at the chessboard and the square, which seemed more appropriate to a follower of al-Khwarizmi than to a poor calligrapher.

After studying both objects carefully, the Man Who Counted spoke as follows: "This interesting square of numbers the calligrapher left behind is what we call a magic square. Let us take a square and divide it up into four or nine or sixteen equal boxes. Put a number in each of these boxes. When the sum of the numbers in any line or column or diagonal always adds up to the same result, we have a magic square. The resulting sum is called the 'constant' of the square, and the number of boxes in any line is the absolute value of the square. The numbers in the separate boxes must all be different. It is impossible to construct a magic square with only four boxes.

"No one knows the origin of magic squares. In times past, making them up was a favorite occupation for the curious. Just as the ancients attributed magical qualities to certain numbers, it was very natural to read magical virtues into these squares. They were known to Chinese mathematicians forty-five centuries before Muhammad. In India, many people used a magic square as a charm. A wise man of Yemen claimed that magic squares could prevent certain diseases. According to

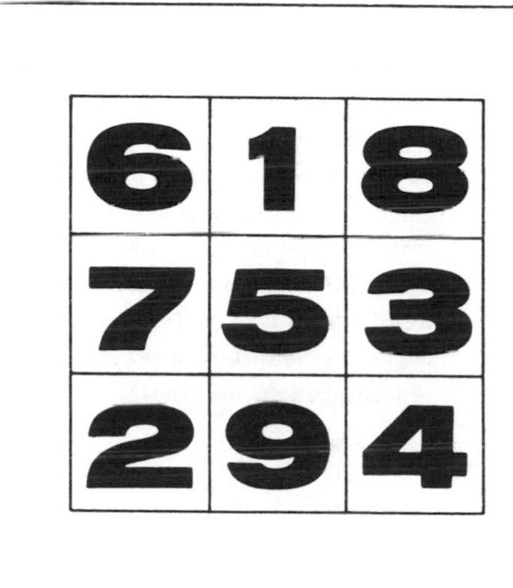


**A magic square of nine numbers.**

certain tribes, a magic square made of silver and worn around the neck was a safeguard against the plague. Ancient Persian magicians, who also practiced medicine, claimed to cure a sickness by applying a magic square, according to the venerable principle *Primum non nocere,* or, the first principle is not to hurt.

"In the field of mathematics, however, magic squares have a curious property. When, for example, a magic square can be broken down into other magic squares, it is called hypermagical. Certain hypermagical squares are known as diabolicals."

Beremiz's observations on magic squares were followed closely by the caliph and his nobles. A wise old man with bright eyes and a squashed nose but with a

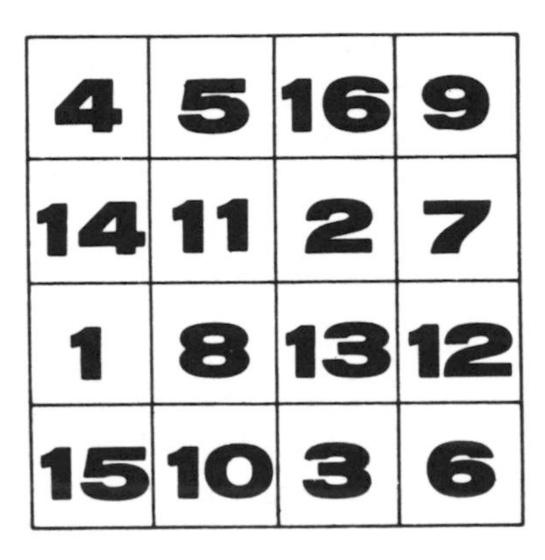

| 4 | 5 | 16 | 9 |
|---|---|---|---|
| 14 | 11 | 2 | 7 |
| 1 | 8 | 13 | 12 |
| 15 | 10 | 3 | 6 |

**This magic square is one that mathematicians call a diabolical. The constant, 34, is obtained not just by adding the numbers in any column, line, or diagonal but also by adding four numbers from the square in many different ways. The corner numbers add up to 34. In fact, there are 86 different ways of reaching the same total.**

warm and merry manner, after first praising "the remarkable Beremiz, the Persian," declared that he wished to seek his opinion.

His question was as follows: "Is it possible for a man versed in geometry to find the exact relation between the circumference and the diameter of a circle?"

The Man Who Counted replied thus: "It is not possible to calculate a circumference exactly, even when we know the diameter. There should be an exact number, but its precise value is unknown to geometers. Ancient astrologers believed that the circumference of a circle was three times its diameter, but that was not so. Archimedes the Greek found that if the measure of a circumference was 22 cubits, its diameter must measure approximately 7 cubits. So, the number sought should be 22 divided by 7. Hindu mathematicians do not agree, and the great al-Khwarizmi declared that Archimedes' law was very far from being true in real life."

And Beremiz, addressing the sage with the squashed nose, concluded, "That very number seems wrapped in mystery, containing qualities that Allah alone can reveal."

Then the Man Who Counted picked up the chessboard and, addressing the king, spoke, "This old board, divided into sixty-four black and white squares, is used, as you know, for an interesting game invented many centuries ago by a Hindu called Lahur Sessa to entertain an Indian king. The discovery of chess is bound up with a legend that involves numbers, calculations, and notable instruction."

"I should like to hear of it," the caliph declared. "It must be interesting."

"I hear and obey," Beremiz replied.

And he told the story recounted in the following chapter.

16
The game plan

*Wherein is told the famous legend of the origin of chess, which Beremiz Samir, the Man Who Counted, tells to the caliph of Baghdad, al-Mutasim Billah, emir of all believers.*

.............................................................................

"It would be difficult to learn, given the uncertainty of ancient documents, the precise period when there lived and reigned in India a prince named Iadava, lord of the province of Taligana. Nevertheless, it would be fair to say that this monarch is considered by various Hindu historians to be one of the richest and most generous sovereigns of his day.

"War, with its fatal procession of calamities, ruined the life of King Iadava, engulfing his royal pleasures with lamentable tribulations. Because it was his royal duty to protect the tranquillity of his subjects, our good and generous monarch was forced to seize the sword of war and, riding at the head of his small army, repulse the sudden and brutal attack of the adventurer Varangul, who was known as the prince of Calian.

"The violent clash of the rival forces covered the fields of Dacsina with the bodies of the dead and made the holy waters of the Sabdhu River run red with blood. According to historians, King Iadava possessed a rare military talent. Before the invasion, he calmly planned his strategy; he then carried it out so skillfully that he totally destroyed the treacherous trespassers who would destroy the peace of his kingdom.

"Unfortunately, his triumph over the fanatic Varangul required many hard sacrifices. Many young soldiers paid with their lives for the safety of the throne and the dynasty. Among the dead left on the battlefield, his chest pierced by an arrow, was Prince Adjamir, son of King Iadava, who had sacrificed himself at the height of the battle to hold the position that secured victory for his country.

"When the bloody campaign was over and the boundaries of his kingdom secure, the king returned to his sumptuous palace in Andra. However, he strictly forbade the traditional, noisy demonstrations with which Hindus typically celebrate their victories. After retiring to his private rooms, he would emerge to consult with his ministers or Brahmin sages only when a grave problem required a decision in the interest of his subjects' welfare.

"With the passage of time, the memories of the painful campaign had not died but grown much worse, plunging the king into agony and sorrow. What good are your rich palaces, your war elephants, the immense treasures you possess, if you have lost the one thing that made life worthwhile? What value could material wealth have in the eyes of an inconsolable father who can never forget his lost son?

"The king could not put out of his mind the ebb and flow of the battle in which Adjamir had died. The unfortunate monarch passed hour after hour tracing in a large sandbox the movements of his troops during the assault. One furrow indicated the advance of the infantry; on the other side, parallel, another line traced the advance of the war elephants. A little beneath, in a symmetrical pattern of circles, was the cavalry under the command of an old captain who was said to enjoy the protection of Techandra, goddess of the moon. In the middle, the king sketched the position of the enemy columns, so badly positioned on the field as a result of the king's own strategy that they were easily, decisively defeated.

"Having completed this sketch of the battle with all the details that he could remember, the king erased it all and started again, as if he gained a certain pleasure in reliving the past in all its private agony.

"In the early morning, when the old Brahmins arrived at the palace to hear the reading of the Vedas, the king had already traced and erased in the box of sand the order of battle that he could not forget or stop retracing.

" 'Unhappy king,' murmured his distressed priests. 'He works like a slave whom God has deprived of reason. Only Dhanoutara, strong and merciful, can save him.'

"And the Brahmins prayed for their king, burning aromatic roots, begging the eternal guardian of the sick to help the king of Taligana.

"Finally, one day, the king was informed that a young Brahmin, humble and poor, was asking for an audience. He had already asked several times, but the king had always refused, claiming that his spirit was not strong enough to receive visitors. But this time he agreed to the request and ordered that the young stranger be brought into his presence.

"After arriving at the throne room, the young Brahmin was questioned, according to the ritual, by one of the king's noblemen.

" 'Who are you? Where do you come from? What do you wish of him who, by the will of Vishnu, is king and lord of Taligana?"

" 'My name,' replied the young Brahmin, 'is Lahur Sessa, and I come from the village of Namir, a thirty days' walk from this handsome city. Word reached the place where I live that our king is afflicted with a deep sorrow, that he is embittered by the loss of the son who was snatched from him by the fortunes of war. It would be a terrible thing, I thought, if our noble sovereign shut himself off in his palace, like a blind Brahmin surrendering to his own pain. Therefore, I thought it might be useful to invent a game that could distract him and open his heart to new pleasures. This is the humble gift that I bring to our King Iadava.'

"Like all the great princes of this or any other book of history, the Hindu sovereign was afflicted by a great curiosity. When he learned that the young Brahmin was offering him a new, unknown game, the king was eager to see and appraise this gift, without delay.

"Sessa brought before King Iadava a board divided into sixty-four squares of equal size. On the board were placed, without obvious care or planning, two sets of game pieces, one white and the other black. The shaped figures were symmetrically placed on the board, and there were curious rules governing their movements.

"Sessa patiently explained to the king, the nobles, and the courtiers who gathered around, the purpose and essential rules of the game:

" 'Each player has eight small pieces, called pawns. They represent the infantry that one sends to thwart the enemy. Supporting the advance of the pawns are the elephants of war, represented by larger, more powerful game pieces. The cavalry, absolutely essential in combat, also appears in the game, in the form of two pieces that can jump over the others, like horses. And to strengthen the attack, there are two lords of the king, two noble and respected warriors. Another piece symbolizes the patriotic spirit of the people and is called the queen. This piece can make many movements and is more efficient and powerful than the rest. Completing the set is a piece that alone can do little but becomes very strong when it is supported by the others. That piece is the king.'

"King Iadava was so interested in the rules of the game that he pursued the inventor with questions:

" 'And why is the queen stronger than the king?'

" 'She is stronger,' explained Sessa, 'because in this game the queen represents the spirit of the people. The throne's greatest strength rests in the exaltation of its subjects. How could the king repulse the enemy's attack without a spirit and commitment to sacrifice in those around him, a spirit that protects the sovereignty of the nation?'

"After a few hours, the king, who quickly learned all the rules of the game, defeated his nobles in a beautifully played game.

"From time to time, Sessa intervened respectfully to clarify a problem or to suggest another plan of attack or defense.

"At one moment, the king noted with great surprise that the pieces, after all the various moves, were deployed in the exact positions of the battle of Dacsina.

" 'Observe,' said the young Brahmin, 'that in order to win this battle this noble must be sacrificed . . .'

"And he indicated precisely the piece that the king had placed at the head of the force fighting in the thick of the battle. The wise Sessa thus demonstrated that at times the death of a prince was required to ensure the peace and liberty of his people.

"On hearing these words, King Iadava, his spirit overcome with enthusiasm, said, 'I did not believe that a clever human could create something as interesting and instructive as this game! By moving these simple pieces, I have just learned that a king is worthless without the support and dedication of his subjects and that, at times, the sacrifice of but one pawn is worth as much as the sacrifice of a powerful piece in winning a great victory.'

"And turning to the young Brahmin, he said, 'I would like to pay you, my friend, for this marvelous gift, which has done so much to relieve my former agonies. Therefore, tell me what you desire, within the realm of what I can give, so that I can show you how grateful I can be to those deserving reward.'

"Sessa appeared unaffected by the king's generous offer. His serene face indicated neither excitement nor joy nor surprise. The apparent indifference of the young Brahmin astonished the courtiers.

" 'Magnificent lord!' the young man replied with both moderation and pride. 'I desire no greater reward for the gift I have brought you than the satisfaction of knowing that I have relieved the lord of Taligana of his infinite sadness. Thus I have already been rewarded, and any other prize would be excessive.'

"The good king smiled somewhat disdainfully at this answer, as it suggested an indifference exceedingly rare among the normally greedy Hindus. Not able to believe in the sincerity of the young man's reply, the king insisted, 'Your disdain and indifference toward material things surprises me, young man. Modesty, when excessive, is like the breeze that extinguishes the light and blinds the old man in the long darkness of the night. So that man can overcome the obstacles that life places in his path, he must subordinate his spirit to an ambition that leads him to a fixed goal. Therefore, you should not hesitate to choose a reward commensurate with the value of your gift to me. Would you like a sack of gold? Would you like a chest of jewels? What about a palace? Would you accept a province of your own to govern? Take care with your answer, and you shall have your reward, by my word of honor!'

" 'After what you have said, a rejection of your offer would be more disobedience than discourtesy,' Sessa replied. 'Thus I will accept a reward for the game I have invented. The reward should fit your generosity. Nevertheless, I do not wish either gold or lands or palaces. I want my reward in grains of wheat.'

" 'Grains of wheat?' exclaimed the king, not hiding his surprise at such an astonishing request. 'How could I reward you in such insignificant currency?'

" 'Nothing could be simpler,' Sessa explained. 'You give me one grain of wheat for the first square on the board, two for the second, four for the third, eight for the fourth, and so on, doubling the amount with each square up to the sixty-fourth and last square on the board. I beg you, O King, in accordance with your magnanimous offer, to pay me in grains of wheat in the manner I have indicated.'

"Not only the king but all the nobles, Brahmins, and everyone present burst out laughing at such a strange request. In fact, the young man's request amazed all those more attached to the material things of life than Sessa was. The young Brahmin, who could have acquired a province or a palace, wished only for a few grains of wheat.

" 'Fool!' exclaimed the king. 'Where did you learn such ridiculous disregard for wealth? The reward you request is absurd. You must know that a handful of wheat contains innumerable grains. With just a few handfuls, I could pay what you ask and more—following your formula of doubling the number of grains with each square on the board. This reward you claim would not even satisfy the hunger of the least village in my kingdom for more than a few days. But, all right, I gave my word, and I shall give you exactly what you have asked for.'

"The king ordered that the most gifted mathematicians of his court be brought before him, and he bade them calculate the amount of wheat due the young Sessa. After a few hours of deep study, the wise calculators returned to the throne room to present the king with their finished calculations.

"Interrupting the chess game he was playing, the king asked the mathematicians, 'How many grains of wheat must I give the young Sessa in order to comply with his request?'

" 'Magnanimous King!' declared the wisest of the mathematicians. 'We have calculated the number of grains of wheat, and we have reached a sum that is beyond human imagination. With the greatest care, we have calculated the number of *ceiras* required to hold the appropriate quantity of wheat, and we have arrived at the following conclusion: the wheat that you will have to give to Lahur Sessa is the equivalent of a mountain with a diameter at its base the size of the city of Taligana and a height ten times greater than that of the Himalayas. If all the fields of India were sown with wheat, in two thousand centuries you would not harvest what you have promised young Sessa.'

"How can one describe the astonishment of the king and his illustrious nobles? The Hindu sovereign realized, perhaps for the first time in his life, that he could not possibly fulfill his promise.

"According to the historians of that period, Lahur Sessa, like a good subject, had no desire to bring grief to his sovereign. After publicly renouncing his request, thus liberating the king from his royal duty, he respectfully spoke to his sovereign:

" 'Consider, O King, the great truth that the wise Brahmins repeat over and over again: the most intelligent men are sometimes blinded not only by the deceptive appearance of numbers but also by the false modesty of the truly ambitious. Unhappy is he who assumes the burden of a debt whose worth cannot be measured by the simple means of his own intelligence. Much wiser is he who praises much and promises little!'

"And, after a short pause, he continued, 'We learn less from the vain science of the Brahmins than by direct experience of life, whose lessons are so frequently underappreciated! The longer a man lives, the more he is beset by moral troubles. One moment sad, the next happy, today fevered, tomorrow tepid, now ambitious, now lazy—so one's composure varies. Only the truly wise man, learned in the laws of the spiritual, can raise himself above these troubles and the fickle whims of mood.'

"These words, so unexpected and so wise, had a profound effect on the king's spirit. Forgetting the mountain of wheat he had promised the young Brahmin, the king named Sessa his first noble.

"And Lahur Sessa, entertaining the king with clever games of chess and instructing him with wise and prudent counsel, poured his blessings on the people and their king, to the greater security of the throne and the greater glory of the nation."

Beremiz's story of the origin of chess enchanted the caliph al-Mutasim. He called the chief scribe and ordered that the legend of Sessa be written on special sheets of cotton paper and stored in a chest of silver.

Then the generous sovereign considered whether he should give the Man Who

Counted a cloak of honor or one hundred pieces of gold.

"God speaks to the world through the hands of the generous."

Such generosity on the part of the ruler of Baghdad delighted everyone. The courtiers in the room were friends of the vizier Maluf and the poet Iezid. They listened with sympathy and accord to the words of the Man Who Counted.

Beremiz, after thanking the sovereign for the presents he had given him, withdrew from the room. The caliph turned to the consideration of his various duties, listening to his ministers of justice and passing wise judgment.

We left the palace at dusk. It was beginning of the month of Shaban.

# 17

# Of apples and ants

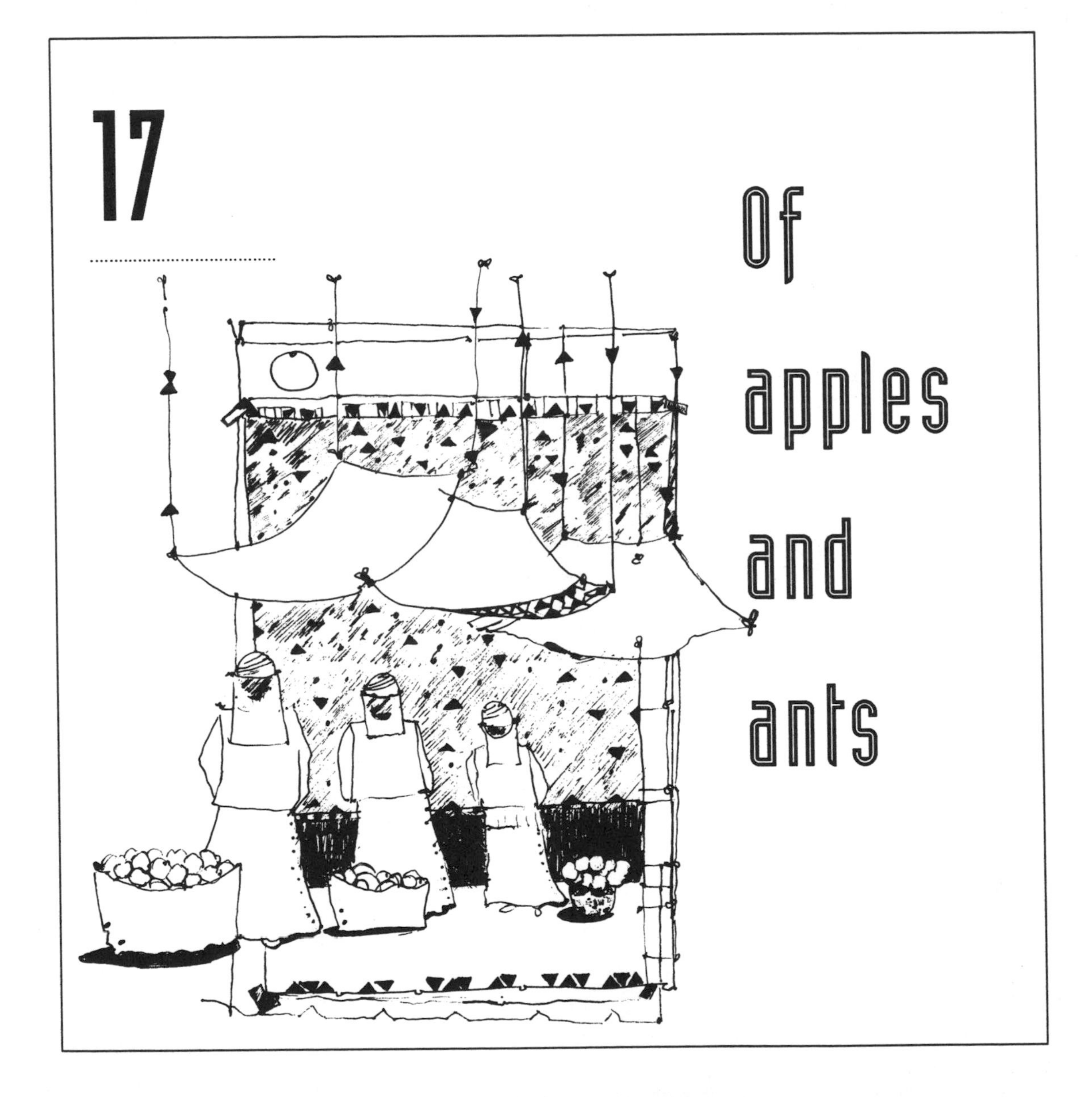

*The Man Who Counted receives many inquiries. Beliefs and superstitions. Numbers and figures. The historian and the calculator. The case of the ninety apples. Science and charity.*

After the famous day when we first sat with the caliph in his throne room, our life changed completely. Beremiz's fame grew extraordinarily. In the modest inn where we lived, the guests showered us with greetings and shows of respect.

Every day the Man Who Counted received dozens of inquiries. A tax collector wanted to know how many *ratls* there were in an *abas,* and how many of both there were in a *cate.* Then a doctor wanted Beremiz to explain how to cure fevers with a cord knotted seven times. More than once, camel drivers and incense sellers came asking the counting man how many times a man had to jump over a fire to exorcise a demon. At times, at dusk, Turkish soldiers with stern faces came to Beremiz and asked for sure ways to win at various games of chance. On other occasions, women, hidden behind heavy veils, came to consult with the mathematician, asking which numbers they should write on their left forearm for good luck, joy, or wealth.

Beremiz Samir received them all with patience and kindness. He gave explanations to some and advice to others. He tried to dispel the superstitions of the ignorant and demonstrated that, according to the will of God, there is no relation between numbers and the joys, sorrows, and anxieties of the heart.

In all his consultations, he was guided only by a sense of altruism, with no expectation of reward. He rejected any money that was offered him, and when a rich sheik whose problems Beremiz had solved insisted on paying him, Beremiz took the sack of money, thanked the sheik, and ordered that the money be distributed among the poor of the district.

Once a merchant named Aziz Neman came, clutching a paper filled with numbers and complaining about a business partner whom he called "a miserable thief," a "hideous jackal," and other epithets equally insulting. Beremiz calmed the man.

"You should beware of reckless judgments," he said, "for these often blind you to the truth. He who looks through a colored glass sees everything in the color of the glass. If the glass is red, all appears bloodied. If the glass is yellow, all appears honeyed. Passion is like a glass before the eyes of man. If someone pleases us, we are all praise and forgiveness. If someone displeases us, however, we judge everything he does harshly."

Then he patiently proceeded to examine the accounts on the paper and found there various errors that made the sums incorrect. Aziz realized that he had been unfair to his partner and was so pleased with the intelligent, thoughtful manner of Beremiz that he invited us to stroll through the city with him that night.

Our companion took us to the Café Bazarique in the Plaza of Ottoman. There a famous historian, sitting in a room filled with heavy smoke, was captivating the patrons with his tales.

We were fortunate to arrive just at the moment that Sheik el-Medah was finishing his introductory speech and beginning his story. He was a man of about fifty, with a very dark complexion, jet black beard, and sparkling eyes. Like all the tale-tellers of Baghdad, he was wearing a large white cloth tied around his head with a cord of camel's hair, and this garb gave him the majesty of an ancient priest. Sitting in the midst of his rapt audience, he spoke in a loud, quavering voice to the accompaniment of a lute and a drum. The men in the café hung on his every word. The sheik's gestures were so dramatic, his voice so eloquent, his face so expressive that at times it seemed that he had actually experienced the adventures that he invented. When he spoke of a long journey, he affected the slow rhythm of a tired camel. At other times, he took on the deep fatigue of a Bedouin searching for a drop of water. Sometimes he would let his head and shoulders droop as if he were a man completely sunk in despair.

In the words of the tale-teller, one could see Arabs, Armenians, Egyptians, Persians, and bronze nomads of Hejaz. How richly he deserved our admiration, this clever, intelligent man who, with soulful eyes, projected the deep sentiments hidden behind the savage visage! The teller of tales moved from one side to the other, then back to the center, then covered his face with his hands, and then threw his arms to the sky, and, just at the moment that he tore the air with his words, the musicians raised a bright flurry of thundering sound.

When he finished, the applause was deafening. Then the customers began to speak among themselves about the most dramatic episodes in the teller's tale.

The merchant Aziz Neman, who seemed to be very popular in this noisy company, moved to the center of the room and spoke to the sheik in a deliberate and solemn tone: "We have among us tonight, O brother Arabs, the celebrated Beremiz Samir, the Persian calculator, secretary of the Vizier Maluf."

Hundreds of eyes turned to Beremiz, whose presence did honor to the patrons of this café.

The storyteller, after respectfully saluting the Man Who Counted, spoke in a clear, well-modulated voice: "Friends! I have told many marvelous stories of kings and genies, good and bad. In honor of the illustrious calculator who joins us this evening, I am going to tell a story that contains a problem to which a solution has never been found."

"Very good, very good!" shouted the audience.

After evoking the name of Allah—to whom all praise and glory!—the sheik told his story.

"There once lived in Damascus an enterprising peasant who had three daughters. One day, the peasant told a *qadi,* a judge, that his daughters were not only very intelligent but blessed with rare skills of the imagination. The *qadi,* a jealous and stingy man, was annoyed at hearing a peasant speak so praisingly of his daughters' talents. In response he declared, 'This is the fifth time that you have told me, in such exaggerated phrases, how wise your daughters are. I am going to summon them to my chambers to see for myself whether they are as clever as you claim they are.'

"The *qadi* had the three girls brought before him. Then he said to them, 'Here are 90 apples for you to sell in the market. Fatima, the oldest, you will take 50; Cunda, you will take 30; and Shia, the youngest, you will take 10. If Fatima sells the apples at a price of 7 to the dinar, you other two will have to sell yours at the same price. And if Fatima sells her apples for 3 dinars each, you two will have to do the same. But, no matter what you do, each of you must end up with the same amount of money from your different numbers of apples.'

" 'But can I not give away some of the apples that I have?' Fatima asked.

" 'Under no circumstances,' said the wretched *qadi*. 'These are the terms: Fatima must sell 50 apples. Cunda must sell 30 apples. And Shia must sell the 10 apples that remain. And all of you must sell your apples at the same price, and all of you must earn exactly the same profit in the end.'

"The dilemma of the three sisters was, of course, an absurd one. How could they possibly solve it? Although they all had to sell their apples at the same price, the sale of 50 apples would vastly exceed the sale of 30 or 10 apples.

"Because the girls did not know how to resolve the problem, they went to a holy man who lived in their district. After filling many pages with calculations, the holy man concluded thus:

" 'Girls, the solution is crystal clear. Sell the 90 apples as the *qadi* has ordered, and you will each receive the same profit.'

"The holy man gave the three sisters directions that did not seem to resolve the problem of the 90 apples at all. But the sisters went to the market and sold their apples as instructed. That is, Fatima sold 50, Cunda sold 30, and Shia sold 10, all at the same price—and each received exactly the same amount as profit. That is the end of the story. I leave it to the calculator to explain how the holy man solved the problem."

Hardly had the storyteller finished than Beremiz spoke to the assembled customers.

"Problems presented in the form of a story are always extremely interesting because the story so beautifully disguises the real problems of mathematical logic. The solution to the problem with which the *qadi* of Damascus tormented the three sisters is the following:

"Fatima starts selling her apples at a price of 7 apples for 1 dinar. She sells 49 of her apples at this price, but keeps back 1.

"Cunda sells 28 of her apples at this price, but keeps back 2.

"Shia sells 7 of her apples at this price, but keeps back 3.

"Then Fatima sells her 1 remaining apple for 3 dinars. In accordance with the rules of the *qadi,* Cunda then sells her 2 remaining apples for 3 dinars each. And Cunda then sells her 3 remaining apples for 3 dinars each.

"Thus:

| | |
|---|---|
| FATIMA | |
| First phase: | 49 apples bring a profit of 7 dinars |
| Second phase: | 1 apple brings a profit of 3 dinars |
| Total: | 50 apples bring a profit of 10 dinars |
| CUNDA | |
| First phase: | 28 apples bring a profit of 4 dinars |
| Second phase: | 2 apples bring a profit of 6 dinars |
| Total: | 30 apples bring a profit of 10 dinars |
| SHIA | |
| First phase: | 7 apples bring a profit of 1 dinar |
| Second phase: | 3 apples bring a profit of 9 dinars |
| Total: | 10 apples bring a profit of 10 dinars |

"Therefore, each made a profit of 10 dinars, and thus the problem set by the envious *qadi* of Damascus was solved."

May Allah condemn the wretched and reward the good!

Then Sheik el-Medah, overjoyed with the solution Beremiz provided, exclaimed, raising his hands to heaven, "By the Second Coming of Muhammad! This young counting man is truly a genius! This is the first man I have met who, without resorting to complicated explanations, perfectly solved the problem of the *qadi*!"

As one, the multitude gathered in the café joined in the sheik's exaltations.

"Bravo! Bravo! Allah praise the wise young man!"

After silencing the noisy crowd, Beremiz continued, "My friends. I must protest that I do not deserve the honorable title of wise man. The man is not wise who merely diminishes ignorance. What value can the science of man have when compared with the science of God?"

And before anyone could reply, Beremiz began the following story:

"There was once an ant that, traveling across the face of the earth, came upon a mountain of sugar. Very happy with his discovery, he removed one grain of sugar from the mountain and took it to his anthill. 'What is this?' his neighbors asked. 'This,' the vain ant replied, 'is a mountain of sugar. I found it in my path and decided to bring it home to you.' "

And Beremiz added, with a fierceness much out of character with his usual tranquillity, "This is the wisdom of the proud—to find a mere crumb and call it the Himalayas. Science is a great mountain of sugar, and from this mountain we satisfy ourselves with mere morsels."

And then he said with great determination, "The only science of value to mankind is the science of God."

A sailor from Yemen asked, "O great calculator, what is the science of God?"

"The science of God is kindness and generosity."

At that moment I remembered the admirable verses spoken by Telassim in the garden of Sheik Iezid when the birds were set free:

> Though I speak with the tongues of men
> and of angels,
> and have not charity,
> I am become as sounding brass,
> or a tinkling cymbal,
>     I am nothing
>     I am nothing

At midnight, when we left the café, several men offered to accompany us with their lanterns through the dark night and the twisted alleys, and soon we were lost. I looked up to the heavens, and there, high in the darkness, shining in the midst of the bright caravan of stars, was the unmistakable star Sirius.

Allah!

# 18

# A perilous pearl

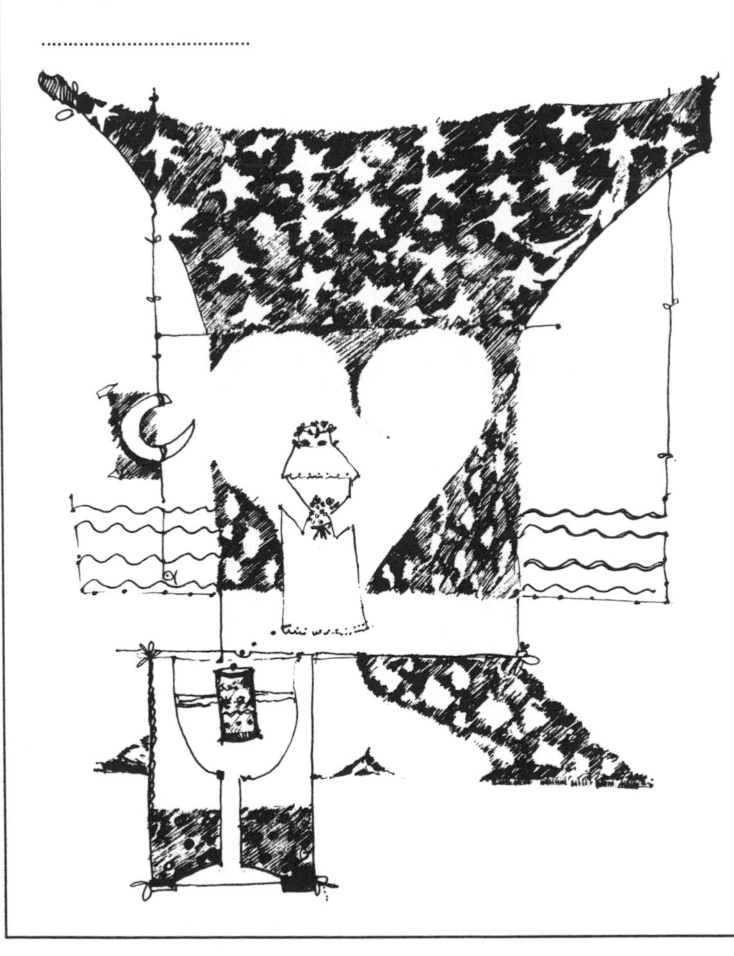

*Which tells of our return to Sheik Iezid's palace. A meeting of poets and learned men. An homage to the maharaja of Lahore. Mathematics in India. The appealing legend of "The Pearl of Lilivati." The great treatises that Hindus wrote on mathematics.*

The following day, at an early hour, an Egyptian arrived at our humble inn with a letter from the poet Iezid.

"It is yet very early for giving classes," said Beremiz patiently. "I'm afraid my pupil may not be prepared."

The Egyptian explained to us that the sheik, before the mathematics lesson, wished to introduce the Persian to a group of friends, and so it would be convenient for us to arrive as early as possible at his palace.

This time, as a precaution, we took with us three black slaves, strong, determined men, for it was very possible that the fearsome and envious Tara Tir might attack us on the way, in an attempt to kill Beremiz, whom he saw as his hated rival.

Nothing untoward happened, and an hour later we arrived at the sumptuous dwelling of Sheik Iezid. The Egyptian servant showed us through an endless gallery into a rich reception room, blue, with gold friezes. We followed him in silence, not without some disquiet on my part at the suddenness of our summons.

There we found the father of Telassim, surrounded by some poets and learned men.

"Peace be with you!"

We exchanged greetings. The master of the house addressed us in the friendliest fashion and bade us be seated. We disposed ourselves on soft silken cushions, and a black slave brought us fruit, pastries, and rose water.

I realized that one of the guests seemed to be a foreigner and was dressed with exceptional luxury. He wore a tunic of white Genoa silk, tied with a blue sash studded with gems, and a splendid dagger encrusted with sapphires and lapis lazuli. His turban was of pink silk decorated with black threads and precious stones. The glitter of fine rings on his delicate fingers highlighted the olive skin of his hand.

"Noble man of numbers," said Sheik Iezid, addressing Beremiz. "I know you must be surprised by this gathering I have summoned to my modest house. I must tell you, however, that its purpose is to pay homage to our illustrious guest, Prince Cluzir el din Mubarek Shah, lord of Lahore and Delhi."

Beremiz, inclining his head, greeted the young man in the jeweled sash, who was the maharaja.

We were well aware, from the gossip of foreigners at our inn, that the prince had left his rich domain in India to fulfill one of the obligations of a good Muslim—namely, to make a pilgrimage to Mecca, the Pearl of Islam. He was spending a few days in Baghdad, and would leave very soon for the Holy City with his countless servants and aides.

"We are very eager," Sheik Iezid went on, "that you help us clear up a question raised by Prince Cluzir Shah. What have Hindus contributed to the advancement of mathematics, and who are the Indian geometers who have made outstanding contributions to this study?"

"Generous Sheik!" replied Beremiz. "I feel that the task you present me with requires both learning and serenity—learning, to know in detail the history of science, and serenity, to analyze and assess it with some discerning. However, O Sheik, your least desires are as commands to me. I shall recount to this distinguished company, as small homage to Prince Cluzir Shah, what little I have learned about the development of mathematics in the country of the Ganges."

And so he began, "Nine or ten centuries before Muhammad, there lived in India a distinguished Brahmin known as Apastamba. This sage, in order to instruct priests on the building of altars and the design of temples, wrote a work called the *Sulbasutra,* which contains numerous mathematical examples. It is most unlikely that this work was influenced by Pythagorean theory, since the Hindu sage does not follow Greek methods of investigation. In his pages, however, there are various theorems and rules for their construction. To illustrate the building of an altar, Apastamba proposes drawing a right-angled triangle, whose sides measure 39, 36, and 15 inches respectively. To resolve the problem, he applies a principle attributed to Pythagoras the Greek:

The area of a square drawn on the hypotenuse is equal to the sum of squares drawn on the two adjacent sides."

And, turning to Sheik Iezid, who was listening attentively, Beremiz spoke as follows: "It would be easier to explain this well-known proposition by a diagram."

Sheik Iezid signaled to his servants, and in a moment two slaves brought in a large sandbox, on the smooth surface of which Beremiz was able to draw figures and outline his calculations for the prince of Lahore. Beremiz drew his diagram with a bamboo stick.

"Here is a right-angled triangle. Its longest side is called the hypotenuse. Let us now draw a square on each of the three sides. Thus it will be easy to prove that the large square drawn on the hypotenuse is exactly equal in its area to the sum of the other two squares, so proving the truth of Pythagoras' principle.

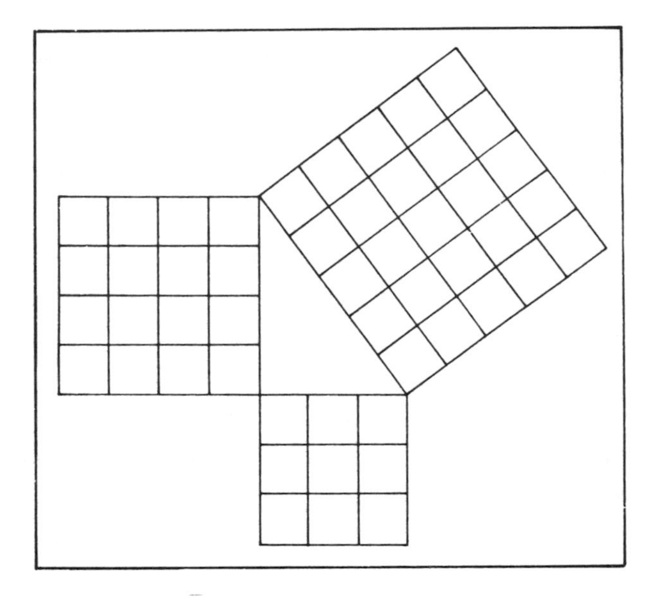

The Prince asked whether the same principle was valid for all triangles.

Beremiz replied gravely, "It is true and constant for all right-angled triangles. I will say, without fear of error, that Pythagoras' law expresses an eternal truth. Even before the sun shone on us, even before there was air to breathe, the square of the hypotenuse was equal to the sum of the squares on the other two sides."

Fascinated by Beremiz's explanation, the prince addressed the poet Iezid warmly: "How wonderful is geometry, my friend! What a remarkable science! In its lessons we perceive two things that move even the humblest and most unthinking of men—clarity and simplicity."

And, lightly touching Beremiz on the shoulder with his left hand, he asked him, "And does this Greek proposition appear in the *Sulbasutra* of Apastamba?"

Beremiz did not hesitate.

"Oh yes, my Prince!" he said. "You will find the so-called Pythagorean theorem in the *Sulbasutra* in a somewhat different form. It was through reading Apastamba that priests learned how to construct oratories, turning right angles into their equivalent squares."

And did other notable works of calculation come out of India?

"Quite a number," Beremiz replied. "I would mention the strange work *Suna Sidauta,* remarkable although by an unknown author. It lays down very simply the laws governing decimals, and shows the supreme importance of zero to the calculator. No less important were the writings of two Brahmin sages much admired today by men of numbers: Aryabhata and Brahmagupta. Aryabhata's treatise is in four parts: on celestial harmonies, on time and its measures, on spheres, and on the elements of calculating. Errors in Aryabhata's writings were not rare—they teach, for example, that the volume of a pyramid can be found by multiplying half the base by its height."

"And that is not true?" asked the prince.

"Completely erroneous," Beremiz replied. "To discover the volume of a pyramid, we must multiply not a half but a third of the base area by its height."

At the prince's side sat a tall, thin man, richly dressed, with red hair showing in his gray beard, who did not by his appearance seem Hindu. I thought he was a tiger hunter, but I was mistaken. He was a Hindu astrologer who was accompanying the prince on his pilgrimage to Meccas. He wore a blue turban, wound around three times, somewhat ostentatiously. He was called Sadhu Gang, and he seemed extremely interested in Beremiz's words.

At one point Sadhu, the astrologer, decided to enter the debate. Speaking in an awkward foreign accent, he asked Beremiz, "Is it true that geometry in India was studied by a wise man who knew the secrets of the stars and the deepest mysteries of the heavens?"

After thinking a moment, Beremiz took up his bamboo stick, cleared off the surface of the sandbox, and wrote a single name:

Bhaskara the Learned

And he said solemnly, "That is the name of India's most famous geometer. Bhaskara knew the secrets of the stars and studied the deepest mysteries of the heavens. He was born in Bidom, in the province of Decan, five centuries after Muhammad. His first text was called *Bijaganita.*"

"*Bijaganita?*" said the man in the blue turban. "*Bija* means seed, and *ganita,* in one of our ancient dialects, means 'to count' or 'to measure.' "

"Just so," nodded Beremiz. "The best translation of that title would be 'The Art of Counting Seeds.' Apart from *Bijaganita,* Bhaskara the Learned wrote another famous work, *Lilavati,* which, as we know, was the name of his daughter."

The astrologer in the blue turban broke in, "It is said that there is a legend surrounding Lilavati. Do you know it?"

"Indeed I do," replied Beremiz, "and if our prince agrees, I could tell it . . ."

"With much pleasure," exclaimed the prince of Lahore. "Let us hear the legend of Lilavati! I am sure it will prove most absorbing."

At that moment, at a sign from Sheik Iezid, five or six slaves appeared in the room and handed around to the guests stuffed pheasants, milk cakes, fruit, and refreshments. When we had finished the delicious repast and made our ritual ablutions, the Man Who Counted was asked to proceed with the story.

Beremiz raised his head, looked at all those who were present, and began to speak.

"In the name of Allah, the Wise and Merciful! Bhaskara the Learned, the famous geometer, had a daughter called Lilavati. When she was born, the astrologer consulted the heavens and, by the disposition of the stars, ascertained that she was destined to remain a maid for her whole life, passed over by the love of fine young men. Bhaskara did not accept such a determination of destiny and consulted the most famous astrologers of the day. How could the gentle Lilavati find a husband and make a happy marriage?

"One astrologer advised Bhaskara to take his daughter to the province of Dravira, close to the sea. In Dravira, there was a temple carved out of stone with a Buddha figure holding a star in his hand. Only in Dravira, the astrologer swore, could Lilavati find a husband, but the marriage would be a happy one only if the wedding ceremony took place on a certain day marked on the cylinder of time.

"To her pleasant surprise, Lilavati's hand was sought in marriage by a rich young man, honest, hardworking, and of a high caste. The day was fixed and the hour marked, and friends came together to attend the ceremony.

"Hindus measured and determined the hours of the day with the help of a cylinder placed in a vase filled with water. The cylinder, open at the top, had a small hole in the center of its base. As the water entered slowly through the hole, the cylinder sank in the vase until it filled it completely, at a predetermined hour.

"Bhaskara placed the cylinder of the hours in position with the greatest care and waited for the water to arrive at the level marked. His daughter, impelled by an irresistible curiosity common to women, wished to observe the rising of the water in the cylinder and leaned over it to watch. One of the pearls on her dress came loose and fell into the vase. By mischance, the pearl, forced by the water, blocked the small hole in the cylinder, as the astrologer had foretold. The groom and the guests withdrew to fix, after consulting the stars, another day for the ceremony. Some weeks thereafter, the young Brahmin who had asked Lilavati's hand in marriage disappeared, and Bhaskara's daughter remained a maid forever.

"Realizing that it is fruitless to fight against destiny, the wise Bhaskara said to his daughter, 'I will write a book that will perpetuate your name, and you will live in the memory of men for a much longer time than the life of the children who would have been born from your ill-fated marriage.'

"Bhaskara's book enjoyed great fame, and the name of Lilavati continues to be immortal in the history of mathematics. What mathematicians refer to as Lilavati is a methodical demonstration of decimal enumeration and of the arithmetic operations of whole numbers. It makes a minute study of the four operations—the problems of elevation, of the square, of the cube, and of the extraction of square roots; and it goes on to study the cube root of any number. Then it tackles fractions with the well-known rule of reducing them to a common denominator. In enunciating these problems, Bhaskara used an elegant and even romantic style. Here is one from his book:

> Much-loved Lilavati, whose eyes are softer than the gentle gazelle's, tell me which is the number that comes from multiplying 135 by 12.

"Another interesting problem in the book refers to the calculations surrounding a swarm of bees:

> A fifth part of a swarm of bees came to rest on the flower of Kadamba, a third on the flower of Silinda. Three times the difference between these two numbers flew over a flower of Krutaja, and one bee alone remained in the air, attracted by the perfume of a jasmine and a bloom. Tell me, beautiful girl, how many bees were in the swarm.

The answer to that is 15. Bhaskara showed in his book that the most complex problems can be presented in a lively, even elegant form."

And Beremiz, continually drawing in the sand, showed the prince of Lahore a number of strange problems culled from *Lilavati.*

Unhappy Lilavati!

And saying the name of the unfortunate girl, I remembered these verses of the poet:

> As the ocean surrounds the earth,
> so do you, my lady,
> surround the heart of the world
> with the abyss of your tears.

# 19 Sailor's choice

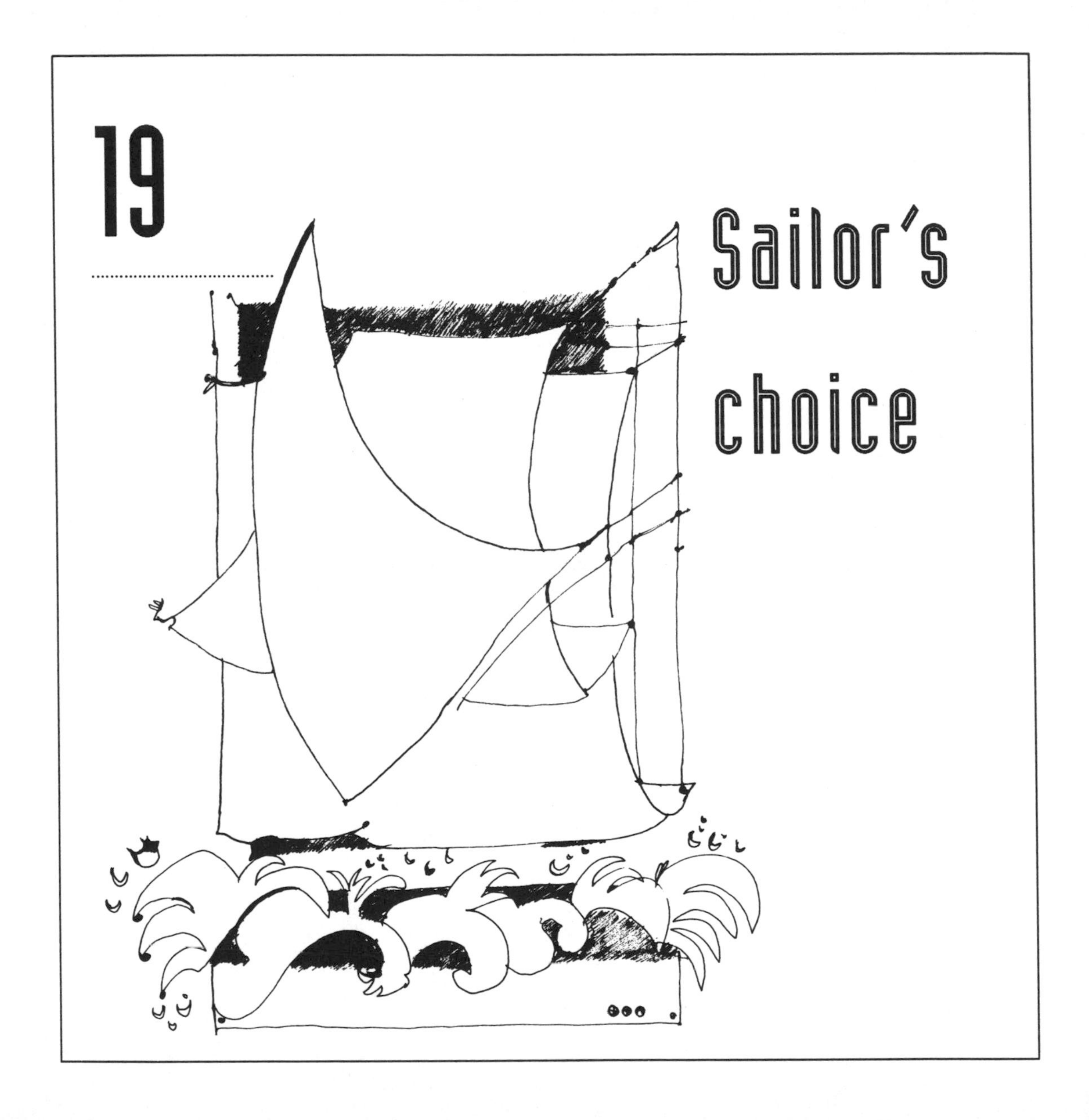

*Prince Cluzir Shah praises the Man Who Counted. Beremiz solves the problem of the three sailors and discovers the secret of the medallion. The generosity of the maharaja of Lahore.*

Beremiz's praise of Hindu science, and its place in the history of mathematics, made a great impression on Prince Cluzir Shah. The young sovereign said that he considered the Man Who Counted a very wise man, one capable of teaching the algebra of Bhaskara to a hundred Brahmins.

"The story of the unfortunate Lilavati, who lost her groom by one pearl of her gown, was enchanting," he added. "The problems of Bhaskara of which the counting man so eloquently spoke indicate the poetic spirit that is so often missing from mathematical works. It is unfortunate, however, that the illustrious mathematician did not mention the famous problem of the three sailors, which is included in so many books and to which no solution has ever been found."

"Magnanimous Prince," replied Beremiz. "I did not mention this problem for the simple reason that I have only a vague knowledge of it and its famous difficulty."

"I know it quite well," said the prince, "and would take great pleasure in repeating this mathematical problem, which has so preoccupied algebraists."

And Prince Cluzir Shah told the following story:

"A ship that was returning from Serendib with a cargo of spices was struck suddenly by a violent storm. The ship would have been destroyed by the furious waves had it not been for the bravery of three sailors who, in the midst of the storm, manipulated the sails with exceptional skill.

"Wishing to reward the brave sailors, the ship's captain gave them a certain quantity of coins. The number was between 200 and 300. The coins were placed in a chest so that, when the ship reached port the following day, the tax collector could divide the sum among the three sailors.

"But during the night one of the three sailors woke and thought, 'It would be better if I took my part right now. That way I would not have to quarrel with my two friends over the money.' Without saying anything to the other two sailors, he got up and found the chest. He divided the money into three parts, but they were not perfectly equal. One coin was left over. 'Because of this one miserable coin,' he thought, 'we will certainly quarrel tomorrow morning. Better to throw it away.' Then he tossed it into the sea and quietly returned to his bed.

"He took his own portion of the reward with him and left the portions of the other two in the chest.

"An hour later, the second sailor had the same idea. He went to the chest and, not knowing that one of his fellows had already withdrawn his portion, he divided the money into three equal parts. Once again, there was one coin left over. In order to avoid any quarreling in the morning, the second sailor did as the first had and tossed the extra coin into the sea. Then he returned to his bed with the portion of money he believed was rightfully his.

"The third sailor had exactly the same idea. Not knowing that his two fellows had already acted, he rose in the early hours of the morning, went to the chest, and divided the money into three parts. Again there was one coin left over after the division, and again he chose to throw it into the sea. Then the third sailor took his one-third portion and happily went back to bed.

"The following morning, when the ship docked, the tax collector found a handful of coins in the chest. He divided the coins into three equal parts and gave one to each of the three sailors. But again the division was not exact. One coin was left over, which the tax collector kept as payment for his services. Of course, none of the three sailors complained about the division, because each believed he had already received his just share.

"So here is the problem: How many coins were in the chest at the beginning? And how many coins did each sailor receive?"

The Man Who Counted, seeing that the prince's story had excited great curiosity among the gathered nobles, decided that he should give a complete explanation of the problem and its solution. Thus he spoke:

"The number of coins which, as you said, were between 200 and 300, must have been 241 before they were divided by the first sailor.

"The first sailor then divided them into three equal parts, throwing one into the sea:

$$241 \div 3 = 80 \text{ coins, leaving } 1$$

"He took his one-third part and then returned to bed, leaving in the chest

$$241 - (80 + 1) = 160 \text{ coins}$$

"The second sailor then divided the 160 coins left in the chest into three parts with one left over, which he threw into the sea:

$$160 \div 3 = 53 \text{ coins, leaving } 1$$

"Then he put his one-third part in his pocket and went back to bed, leaving in the chest

$$160 - (53 + 1) = 106 \text{ coins}$$

"The third sailor then divided the 106 coins into three equal parts with one left over, which he threw into the sea:

$$106 \div 3 = 35 \text{ coins, leaving } 1$$

"Then he returned to bed with his one-third part, leaving in the chest

$$106 - (35 + 1) = 70 \text{ coins}$$

"This was the number of coins left when the ship docked and the tax collector, following the instructions of the ship's captain, divided the coins equally in three parts with one left over.

$$70 \div 3 = 23 \text{ coins, leaving } 1$$

"The tax collector thus gave 23 coins to each sailor and kept the one left over for himself. Therefore, the division of the original 241 coins went like this:

| | | |
|---|---|---|
| *First sailor* | *80 + 23* | *103* |
| *Second sailor* | *53 + 23* | *76* |
| *Third sailor* | *35 + 23* | *58* |
| *Tax collector* | | *1* |
| *The sea* | | *3* |
| *Total* | | *241* |

Having solved the problem, Beremiz fell silent.

The prince of Lahore took a silver medallion from his pocket and, turning to the Man Who Counted, said, "Because of the clear and simple solution you have given to the problem of the three sailors, I see that you might also be capable of explaining even more intricate numerical problems.

"This piece," the prince continued, "was engraved by a religious artist who lived for some time in the court of my grandfather. On the piece, the engraver set a puzzle that, until now, no magician or astrologer has been able to solve. On one side of the medallion, the number 128 is surrounded by seven small rubies. On the other side, divided into four parts, are the numbers

*7, 21, 2, and 98*

"As one can see, the sum of the four numbers is 128. But what is the meaning of the four parts into which the number 128 is divided?"

Beremiz took the medallion from the prince's hand. For a time he examined the medallion in silence and then spoke as follows:

"This medallion, O Prince, was engraved by a man well steeped in numerical mysticism. The ancients believed that certain numbers had magical powers. The number 3 was considered divine and 7 sacred. The seven rubies surrounding the number 128 show that the engraver was preoccupied by the relation between the number 128 and the number 7. The number 128, as we know, can be divided into 7 multiplications of 2:

$$2 \times 2 \times 2 \times 2 \times 2 \times 2 \times 2$$

"The number 128 can also be divided into four parts:

*7, 21, 2, and 98*

which suggests the following properties: add 7 to the first, subtract 7 from the second, multiply the third by 7, divide the fourth by 7, and you will get the same result:

$$7 + 7 = 14$$
$$21 - 7 = 14$$
$$2 \times 7 = 14$$
$$98 \div 7 = 14$$

"This medallion must have been used as a talisman, for it contains relations that depend on the number 7, which was considered sacred."

The prince of Lahore was so enchanted with Beremiz's explanation that he offered, as a reward, not only the medallion but also a sack of gold coins. The prince was generous and good.

Then we filed into a grand salon, where Sheik Iezid offered his guests a splendid banquet. Bit by bit, Beremiz's esteem was growing, proving that he was destined to occupy a distinguished position much greater than one of his poor upbringing could have expected.

Some of the guests could not hide their disappointment. As for me, I was of no consequence.

# 20

# The power of ten

*Beremiz gives his second mathematics lesson. Numbers and the idea of numbers. Figures. Number systems. Decimal numbers. Zero. We hear again the exquisite voice of the invisible pupil. Doreid, the grammarian, quotes a poem.*

The repast concluded, the Man Who Counted rose at a sign from Sheik Iezid. It was the appointed hour for his second mathematics lesson, and his invisible pupil was waiting for her teacher.

After taking his leave of the prince and the attendant sheiks, Beremiz, accompanied by a slave, made his way to the room set aside for the class. I rose also and went with him, since I wanted to take full advantage of the permission that had been granted me to attend the lessons Beremiz was giving to young Telassim.

One of the guests, the grammarian Doreid, a friend of the household, also showed an interest in attending; taking leave of the prince, he followed us. He was middle-aged, of a cheerful disposition, with a fine, expressive face.

We passed through an elegant gallery covered in Persian carpets, and, shown the way by a startlingly beautiful Circassian slave, we reached the room where Beremiz was to give his class. The red carpet that had hidden Telassim some days before had been replaced by a blue one with a starred heptagon in its center.

The grammarian Doreid and I sat in a corner of the room close to a window opening on the garden. Beremiz settled himself, as on the first occasion, in the center of the room on an ample silk cushion. By his side, on a small ebony table, lay a copy of the Koran. The Circassian slave and another Persian with soft, smiling eyes stationed themselves beside the door. The Egyptian slave, Telassim's personal guard, leaned against a column.

After a prayer, Beremiz began.

"We do not know when the idea of numbers first arose. The researches of philosophers go back to a time hidden from us by the clouds of the past.

"Those who study the evolution of numbers show that, even among primitive men, human intelligence had a special faculty that we shall call the sense of number. That faculty allows us to know, in a purely visual way, whether a group of things has grown or diminished—that is, whether it has undergone numerical change.

"The sense of number should not be confused with the ability to count. Human intelligence alone can achieve the level of abstraction that we call the sense of number, while the ability to count can be observed in many animals. Some birds, for example, are able to count the eggs they leave in the nest, distinguishing between two and three, and certain wasps can tell the difference between five and ten.

"Tribesmen of North Africa knew all the colors of the rainbow and named each one, but the tribe had no word for color. In the same way, many primitive languages have words for one, two, three, and so on, but no separate word for number.

"Where does the idea of number come from?

"We do not know how to answer that, my lady. In the desert, a Bedouin sees a caravan in the distance, moving slowly. The camels approach with their burdens of men and goods. How many camels are there? We have to answer that question with a number. Forty, perhaps? A hundred? To give an answer, the Bedouin must do something special; he must 'count.' In order to count, the Bedouin must connect each object in the series with a certain symbol: one, two, three, four, and so forth. To arrive at a result of his counting—or, in other words, a number—the Bedouin needs to invent a 'number system.'

"The oldest number system is the quinary, the system that groups its units in fives. Each group of five units is called a quine. Eight units are one quine plus three, written as 13. It has to be made clear that in this system the figure on the left is worth five times what it would be if it were on the right. As mathematicians put it, this system has a base of 5. There are traces of this system in the poems of antiquity.

"The Chaldeans had a number system with a base of 60. In ancient Babylon, the symbol 1.5 stood for the number 65.

"A number of people also used a system with a base of 20. In this system, our number 90 would be written as 4.1—that is to say, four twenties plus ten.

"Following this, my lady, came the system with a base of 10, much more advantageous for working with greater numbers. The origins of that system come from the number of fingers on two hands. In certain kinds of trading, we find a distinct preference for a base of 12, the system that counts by dozens, half dozens, quarter dozens, and so on. Twelve has a considerable advantage over 10 in that it has more divisors.

"The system with a base of 10, the decimal system, has been universally adopted. From the Tuareg who counts on his fingers to the mathematician with his calculators, all count in tens. Given the profound differences between peoples, such universality is surprising: no religion, moral code, form of government, economic plan, philosophical structure, language, or alphabet can boast of anything like it. Counting is one of the few matters about which men do not differ. They consider it both simple and natural.

"If we take note, my lady, of savage tribes and of the ways of children, it becomes obvious that the fingers are the basis of our number system. Using our ten fingers, we begin to count in tens, and our whole system is founded on groups of ten.

"Very probably, the shepherd who at evening had to be sure that all his sheep had entered the fold had to count beyond the first ten sheep. As the sheep passed him, he counted each on a finger, and, after each ten, he let fall a stone. When his count was complete, the stones stood for the number of complete hands or tens of sheep. On the following day, he could repeat his count by counting the pile of stones. Later on, some mind capable of abstraction discovered that this could equally apply to other useful things like fruit, wheat, days, distances, and stars. And when, instead of using stones, we made distinct and lasting marks, a system of written numbers came into being.

"All peoples used the decimal system in their spoken language. Other systems were forgotten. But the adaptation of that system to written numbers came about very gradually. It took several centuries for humanity to find a perfect solution to the problem of writing down numbers. To stand for numbers, man had to think up special characters called figures or ciphers, each one of which stood for one, two, three, four, five, six, seven, eight, or nine. Other extra characters like *d, c,* and *m* showed that the accompanying number stood for ten, a hundred, a thousand, and so forth. Thus a mathematician of antiquity wrote the number 9,765 as 9m7c6d5. The Phoenicians, the most assiduous traders of antiquity, used accents instead of letters: 9‴7″6′5.

"At first, the Greeks did not use this system. Instead, they gave a value to each letter of the alphabet, adding an accent to it. The first letter, alpha, was 1; the second, beta, was 2; the third, gamma, was 3; and so on up to 19. Six was an exception and had its own sign, sigma. The letters were combined in pairs: 20, 22, and so on.

"In the Greek system, the number 4,004 was represented by two figures, the number 2,022 by three, and the number 3,333 by four distinct figures.

"The Romans showed less imagination, using three characters—I, V, and X—to form the first ten numbers and, in combination with them, L (50), C (100), D (500), and M (1,000). Numbers written in Roman figures are thus absurdly complicated and were most unsuitable for the simplest arithmetic, making even small calculations a torture. Roman numbers could be added, but they had to be written one below the other in such a way that the figures with the same terminal letter fell in the same column, making it necessary to leave spaces between figures.

If thou art idle or uncared for, letting the pitcher float on the water, come, come to me.

The grass is greening on the hill, and the flowers on the woods now are opening.

Thy thoughts shall fly from thy dark eyes as birds fly from their nests, and thy veil shall fall at thy feet.

Come, o come to me!

We departed somewhat wistfully from the light-filled room. I noticed that Beremiz was not wearing on his finger the ring he had won in the inn on the day of our arrival. Could he have lost that fine jewel?

The Circassian slave was looking about vigilantly, as if she feared the spell of some invisible genie.

# 21

# The writing on the wall

*I begin my work of copying texts on medicine. The invisible pupil makes great advances in her studies. Beremiz is called in to resolve a complicated problem. King Mazim and the prisons of Khorasan. Sanadik, the smuggler. A poem, a problem, and a legend. King Mazim's judgment.*

Our life in the splendid city of the caliphs was becoming busier every day. Vizier Maluf set me to copying two books by the philosopher Rhazes, books containing a great deal of medical knowledge. In them I read important observations on the treatment for scarlet fever, the curing of childhood diseases, diseases of the kidneys, and a thousand other maladies that afflict men. Involved as I was in this work, I could no longer attend Beremiz's classes in Sheik Iezid's palace.

From what I heard from my friend, the invisible pupil had made great advances in the last weeks. Now she had mastered the four numerical operations, as well as the first three books of Euclid, and could solve fractions with the denominators 1, 2, and 3.

One day, at the end of the afternoon, we were about to begin our modest meal, consisting of meat pies, honey, and olives, when we heard in the street a great tumult of horses and the shouts, orders, and oaths of Turkish soldiers. Startled, I rose to my feet. What was happening? I had the impression that the inn had been surrounded by troops and that some violent action by the ill-tempered police chief was about to occur. The sudden uproar did not upset Beremiz. Completely oblivious, he went on drawing with a carbon stick geometric figures on a wooden tablet. How extraordinary he was! If Asrail, the angel of death, had suddenly materialized bearing the inevitable sentence, he would have continued impassively drawing curves and angles and studying the properties of figures and numbers.

Into the little inn burst old Salim, accompanied by two black slaves and a camel driver, all of them excited, as if something terrible had happened.

"For Allah's sake!" I cried out impatiently. "Don't disturb Beremiz! What is all this tumult about? Has there been a revolt in Baghdad? Has the Mosque of Suleiman fallen down?"

"Master," stammered old Salim, "an escort of Turkish soldiers has just arrived."

"By Allah, what escort, Salim?"

"The escort of the great Vizier Maluf. The soldiers have orders to bring Beremiz Samir to him immediately."

"But why so much noise?" I exclaimed. "This is not so unusual. Naturally enough, the vizier, our great friend and protector, wishes urgently to solve some mathematical problem and needs the help of our wise friend."

My predictions were as sure as the calculations of Beremiz. Moments later, accompanied by the officers of the escort, we arrived at the palace of Vizier Maluf.

We found the vizier in his reception room, surrounded by three aides. He held in his hand a page filled with numbers and calculations. What new problem could have arisen to upset the caliph's worthy adviser so profoundly?

"This is very serious," said the vizier to Beremiz. "I find myself confronted by one of the most difficult problems I have ever faced in my life. Let me tell you in detail how it arose, for only with your help can we possibly find a solution."

And the vizier told us the following:

"The day before yesterday, a few hours before our noble caliph was to leave for Basra for a stay of three weeks, there was a terrible fire in the prison. The prisoners shut up in their cells underwent a long torment of unspeakable agony. Our generous sovereign immediately decided to cut in half the sentences of all the prisoners. At the beginning, we thought nothing of it, for it appeared quite simple to carry out the king's command to the letter. The following day, however, when the caravan of the prince of believers was already far away, we discovered that his last-minute orders raised an extremely delicate problem, which seemed to have no ideal solution.

"Among those sentenced to prison is a smuggler from Basra, Sanadik, who has already served four years of a life sentence. This man's sentence is to be cut in half. But since he was sentenced to prison for the rest of his life, now under the law his sentence must be reduced to half his remaining life. But we have no idea how long he will live, so how can we divide an incalculable time?"

After some moments of thought, Beremiz spoke, choosing his words with great care.

"Your problem seems to me an extremely delicate one, since it involves both mathematics and the interpretation of the law, and is as much a matter of human justice as of numbers. I cannot apply to it any rigorous analysis until I have visited the cell of the condemned man Sanadik. It may be that the *x,* the unknown in the life of Sanadik, has already been decided by destiny on the wall of his cell.

"What you say seems very strange indeed," remarked the vizier. "I cannot see any connection between the outpourings with which madmen and prisoners cover the prison walls and the solution of so delicate a problem."

"My lord!" exclaimed Beremiz. "On prison walls are to be found many interesting writings, formulas, verses, and inscriptions that affect our spirits and lead us to feelings of mercy. It so happens that on one occasion, King Mazim, ruler of the rich province of Khorasan, was informed that a convict had written magic words on the wall of his cell. King Mazim summoned a diligent scribe and ordered him to copy down all the letters, numbers, verses, and writings that he could find on the gloomy walls of the prison. It took many weeks for him to fulfill the king's strange command. Finally, after working with great patience, he brought him pages covered with symbols, unintelligible words, nonsensical drawings, and blasphemies and meaningless numbers. Was there any way to decipher or translate the incomprehensible writings that filled those pages? One of the country's sages, called in by the king, declared, 'Your Majesty, those pages contain curses, heresies, cabalistic words, legends, even a mathematical problem.'

"The king answered, 'Curses and heresies do not interest me. The words of the Cabala leave me indifferent. I don't believe in the secret power or letters or mysteries concealed in human symbols. I am interested, however, in the poems and the legends, since these are noble utterances in which men can find comfort, lessons for the ignorant, or warnings to the powerful.'

" 'A condemned man's despair does not give rise to inspiration,' said the wise man.

" 'Even so, I want to see these writings,' replied the king.

"So the wise man picked up at random a page that the scribe had copied, and read:

> 'Happiness is difficult since we lack the materials that make us happy.
> Do not speak of happiness to those less fortunate.
> When one does not have that which one loves, one must love that which one has.'

"The king remained silent as if lost in deep thought, and the wise man, to distract him, went on, 'Here is one problem scratched on the cell wall of a condemned man:

> Arrange ten soldiers in five rows in such a way that each row has four soldiers.'

"This problem, which looks impossible, has a very simple solution, as is indicated by this drawing, which shows five rows with four soldiers in each one.

"The wise man continued reading:

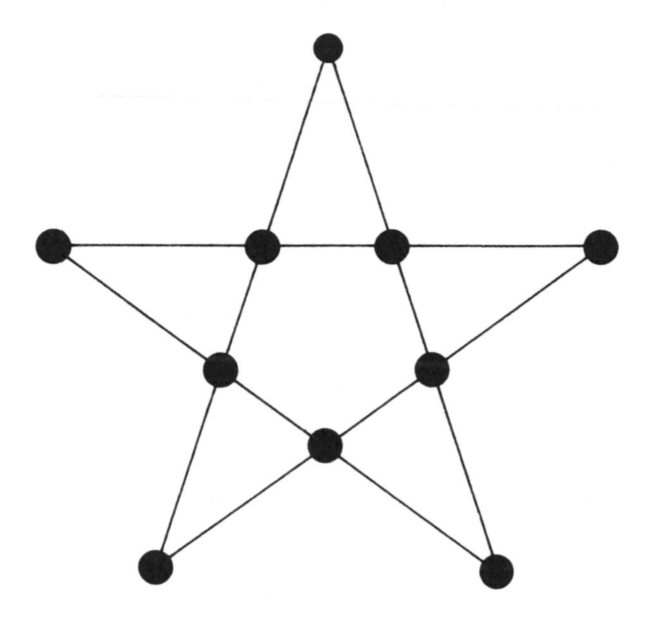

They tell a story of how young Tzu-chang went one day to the great Confucius and asked him, 'How many times, O illustrious one, should a judge reflect before passing sentence?'

"Today once; tomorrow ten times," replied Confucius.

Prince Tzu-chang was startled at hearing these words and did not understand.

'Once should be enough,' explained the master patiently, 'for the judge, after examining the case, to decide on a pardon. A judge,

> however, ought to think ten times before passing a sentence of condemnation.'
>
> And he concluded with this piece of profound wisdom: 'He who hesitates in granting a pardon may be making a grave mistake, but he who condemns without hesitating makes a much greater mistake in the eyes of God.'

"King Mazim was filled with admiration when he realized that on the damp walls of prison cells were to be found such treasures written by wretched prisoners, so many things of beauty and interest. Obviously, among those who watched the bitter days pass from the depths of their cells, there were people of culture and intelligence. So the king ordered all sentences to be revised and discovered that many of them involved cases of blatant injustice. In the light of what the scribe had revealed, innocent prisoners were immediately freed and many judicial errors were corrected."

"Be that as it may," replied Vizier Maluf, "but it is possible that, in the prisons of Baghdad, you will not find geometric figures or moral legends or poems. Nevertheless, I wish to see where your inquiries will lead. And so I give you leave to visit the prison."

# 22

# Half and half

*Of all that happened in the course of our visit to the Baghdad prison. How Beremiz solved the problem of halving the years of life remaining to Sanadik. The instant of time. Conditional liberty. Beremiz explains the basis of a sentence.*

The great prison of Baghdad looked like a Persian or Chinese fortress. Entering, we crossed a small patio in the center of which stood the famous Well of Hope. That was where a condemned man, hearing sentence passed on him, gave up forever all hope of salvation. No one could imagine the suffering and the misery of all those who lived deep in the dungeons of that magnificent Arab city.

The cell of the unfortunate Sanadik lay in the deepest part of the prison. We were led down to the cellars by a jailor and accompanied by two guards. A Nubian slave, a giant of a man, carried the huge torch whose light showed us the recesses of the prison.

Going down a narrow corridor that was barely wide enough for a man to pass, we descended a dark, dank staircase. Deep in the cellars was the small cell where Sanadik was shut up. Not a single ray of light penetrated that darkness. The heavy, fetid air was difficult to breathe without nausea. The floor was covered with a layer of putrid mud, and within the four walls there was not even a bunk for the condemned man to stretch out.

By the light of the torch carried by the great Nubian, we saw the unfortunate Sanadik, half naked, with a thick, tangled beard and long hair falling over his shoulders, crouched on a flagstone, his hands and feet fastened to irons.

Beremiz watched him in an attentive silence. It seemed hard to believe that the unfortunate man could have clung to life throughout four years in that wretched, inhuman situation.

The cell walls, stained and dripping, were covered with writings and with figures, bizarre signs of many generations of prisoners. Beremiz examined them closely, reading and translating with great concentration, stopping every now and again to make long, laborious calculations. How could the counting man decide, on the basis of these curses and blasphemies, how many years of life were left to Sanadik?

Great was my relief at leaving behind the gloomy prison where the wretched prisoners were tortured. When we reached the sumptuous reception room, Vizier Maluf appeared, surrounded by courtiers, secretaries, various sheiks, and wise men of the court. They were all waiting for Beremiz's arrival, wishing to know the formula the Man Who Counted would use to solve the problem of halving a life sentence.

"We await you, O counting man," said the vizier affably, "and I beg you to give us a solution without any delay. We are most eager to comply with the orders of our great emir."

Beremiz bowed respectfully, made his salaams, and spoke as follows:

"Sanadik of Basra, the smuggler, captured four years ago at the frontier, was sentenced to life imprisonment. That sentence has just been cut in half by the wise and just decree of our gracious caliph, ruler of believers, servant of Allah on earth. . . .

"Let us give the value of $x$ to the period of Sanadik's life that begins with his capture and his life sentence. The latter condemned Sanadik to $x$ years in prison, that is, to imprisonment for life. Now, by virtue of the royal decree, that sentence is reduced by half. It is important to state, if we divide the time that $x$ stands for into separate periods, that for each period in prison there must be an equal period of freedom."

"Exactly!" exclaimed the vizier keenly. "I follow your reasoning very well."

"Now, as Sanadik has already served four years, it seems clear that he must have an equal period of freedom, that is, four years. In fact, let us imagine that a friendly magician could foresee the precise number of years left to Sanadik and could say to us, 'This man had only eight years of life left to him when he was imprisoned.' In that case, $x$ would be equal to eight, that is to say, Sanadik had been condemned to eight years in prison, now reduced to four. But since Sanadik has already served four years, he has in fact completed his sentence and must be considered a free man. If the smuggler by the will of destiny were to live more than eight years, if $x$ were greater than eight, his life would be made up of three periods: one of four years in prison already served, another of four years of freedom, and a third, which also would have to be divided into two parts, imprisonment and freedom. This makes it easy to conclude that whatever value $x$, the unknown, may have, the condemned man must immediately be set free to enjoy four years of freedom. He has absolute right to this period of liberty, as I have shown, in accordance with the law.

"Once this period is over, he must return to prison and remain there for a period equal to half the rest of his life. It would be easiest perhaps to imprison him for a year and then grant him freedom for the following year. Thanks to the caliph's command, he would be in prison for one year and free for one year, in that way enjoying the benefits of the caliph's mercy. Such a solution, however, would be a precise one only if the condemned man were to die on the very last day of one of his periods of freedom.

"Let us imagine that Sanadik, after a year in prison, is set free and dies in the fourth month of his freedom. Of that part of his life—one year and four months—he would have spent one year locked up and four months at liberty. That would not be correct, an error of calculation, for his sentence would not have been cut by half.

"It would be simpler perhaps to imprison Sanadik for a month and set him free for the following month. That solution, however, could lead to a similar error if he, after spending a month in prison, did not enjoy a whole month of freedom.

"It seems, you will say, that the best solution in the end would be to imprison Sanadik for one day and set him free the next for an equal period of freedom, continuing thus until the end of his life. Yet that solution would still not satisfy the demands of mathematical exactitude, because Sanadik might die a few hours after a day in prison. Keeping him imprisoned for an hour and then setting him free, and going on in that way until the last hour of his life, would be a solution only if Sanadik were to die in the last minute of one of his hours of freedom.

Otherwise, his sentence would not have been reduced by half, as the caliph's decree commands.

"The exact mathematical solution is as follows: Keep Sanadik in prison for an instant of time and set him free the following instant. It is necessary, however, that his prison time, one instant, be so small as to be indivisible. The same applies to the period of liberty that follows.

"In reality, such a solution is impossible. How can you lock up a man for one indivisible instant and set him free the following instant? So that idea has to be put aside as impossible. I can see, O Vizier, only one way of solving the problem. Let Sanadik be given conditional liberty under the eye of the law. It is the only way of allowing him to serve his sentence and to be free at the same time."

The vizier ordered that Beremiz's suggestion be immediately implemented, and that very day the imprisoned Sanadik was granted conditional liberty, a judgment that Arab magistrates adopted often from then on in passing wise sentence.

The following day I asked Beremiz what details and calculations he had managed to gather from the prison wall during our famous visit, and what had led him to come up with such an original solution to the problem. This was his answer:

"Only someone who has been even for a brief moment within the gloomy walls of a dungeon can know how to solve those problems in which numbers form a terrible part of human misery."

# 23

# All is relative

*What happened in the course of an honored visit to us. The words of Prince Cluzir Shah. A princely invitation. Beremiz solves a new problem. The rajah's pearls. A cabalistic number. Our departure for India is set.*

The humble district in which we lived was enjoying a glorious morning when Beremiz received an unexpected visit from Prince Cluzir Shah. The lavish retinue crowded down our street, bringing the curious out onto the rooftops and balconies. Old men, women, and children gaped at the wonderful spectacle in dumb astonishment. First came thirty horsemen mounted on superb Arab chargers with gold trappings and velvet cloths edged in silver. They wore white turbans and helmets that shone in the sunlight, cloaks and tunics of silk, and scimitars hanging from belts of worked leather. They carried before them standards showing the prince's shield, a white elephant on a blue background. After them came archers and scouts, also mounted.

Bringing up the rear was the all-powerful prince, accompanied by two secretaries, three doctors, and ten pages. He wore a scarlet tunic decorated with rows of pearls. Sapphires and rubies sparkled in his turban. When old Salim saw the magnificent assembly from the inn, he almost went mad. He cast himself to the floor and began to shout, "What is this? Where am I?"

I sent a water carrier to lead my poor old friend to the patio until he recovered his calm. The main room of the inn was too small for the illustrious company. Beremiz, somewhat awed by the gracious visit, went down to the patio to receive his visitors.

Prince Cluzir Shah, entering with his retinue, greeted the Man Who Counted with a friendly salaam and said to him, "A poor wise man is he who seeks out the rich, but a nobler rich man he who seeks out wise men."

"I realize, my lord," replied Beremiz, "that your words are inspired by deep feelings of friendship. The insignificant knowledge that I have acquired is as nothing before your generous heart."

"My visit is dictated more by my own desires than by love of science," replied the prince. "Since I first had the honor of listening to you in the house of Sheik Iezid, I have considered offering you some suitable position in my court. I wish to appoint you my secretary or, even better, master of the Delhi Observatory. Will you accept? We shall leave in a few weeks for Mecca, and from there we shall go on directly to India."

"My generous Prince!" replied Beremiz, "unfortunately I cannot leave Baghdad at this moment. I am bound to this city by a serious obligation. I shall be able to leave only when the daughter of the illustrious sheik Iezid has mastered the beauties of geometry."

The maharajah smiled and replied, "If your refusal is because of that obligation, I think that there is a way of fulfilling it. Sheik Iezid told me that, thanks to her progress, young Telassim in a few months will be ready to teach even wise men the famous problem of the rajah's pearls."

I sensed that our noble visitor's words surprised Beremiz, for he seemed somewhat confused.

"I would be delighted," the prince went on, "to understand this difficult problem, which eludes many calculators and which was first set by one of my distinguished ancestors."

Accordingly, Beremiz in compliance with his wishes spoke on the problem in his slow and steady manner.

"It is less a problem than an arithmetical curiosity," he said. "The situation is as follows: A rajah on his death left to his daughters a certain number of pearls with instructions that they be divided up in the following way: his eldest daughter was to have one pearl and a seventh of those that were left. His second daughter was to have two pearls and a seventh of those that were left. His third daughter was to have three pearls and a seventh of those that were left. And so on. The youngest daughters went before the judge complaining that this complicated system was extremely unfair. The judge, who, as tradition has it, was skilled in solving problems, replied at once that the claimants were mistaken, that the proposed division was just, and that each of the daughters would receive the same number of pearls.

"How many pearls were there? How many daughters had the rajah?

"The solution to this problem is really not at all difficult. Observe.

"There were 36 pearls in all, and the rajah had six daughters. The first daughter received one pearl and one-seventh of the 35 pearls remaining, that is, 5. She received 6 pearls in all, and 30 were left.

"The second daughter received 2 pearls and one-seventh of the 28 remaining pearls, that is, 4. In all, she received 6 pearls, and 24 were left.

"The third daughter received 3 pearls and one-seventh of the 21 remaining pearls, that is, 3. In all, she received 6 pearls, and 18 were left.

"The fourth daughter received four pearls and one-seventh of the 14 that were left, that is, 2. In all, she received 6 pearls, and 12 were left.

"The fifth daughter received 5 pearls and one-seventh of the 7 that were left, that is, 1. She received 6 pearls in all, leaving 6, which became the share of the sixth and youngest daughter."

And Beremiz concluded, "As you see the problem, although ingenious, is not at all difficult. No subtlety is required to reach its solution."

At that moment, the prince's attention was caught by a number written five

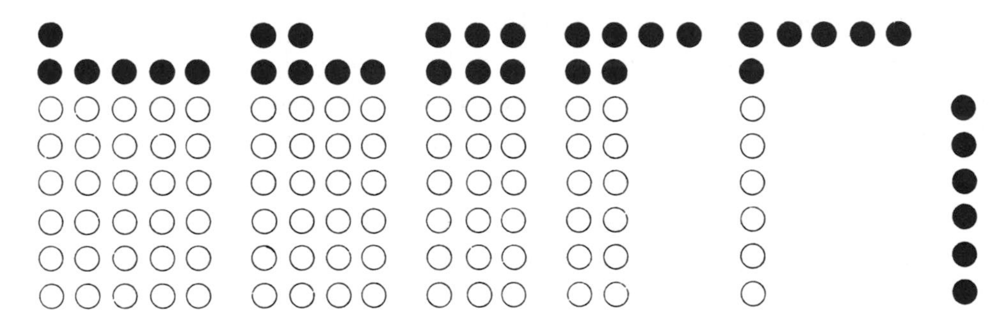

**Graphic solution to the famous problem of the rajah's pearls.**

times on the wall of the room. The number was 142,857.

"What is the significance of that number?" he asked.

"That is one of the oddest numbers in all of mathematics," Beremiz replied. "There are extraordinary coincidences in its relations with its multiples.

"Let us multiply it be two:

$$142{,}857 \times 2 = 285{,}714$$

"Notice that the numbers in the product are the same as those in the given number but in a different order. The 14 that was on the left has moved to the right.

"Let us multiply now by three:

$$142{,}857 \times 3 = 428{,}571$$

"Once more, notice how curious the answer is. Its numbers are the same ones again, but in a different order. The 1 that was at the left has moved to the right; the other numbers stay where they were.

"When we multiply by four, although the numbers shift position, they remain in the same order:

$$142{,}857 \times 4 = 571{,}428$$

"The same thing happens when we multiply it by five:

$$142{,}857 \times 5 = 714{,}285$$

"Watch what happens when we multiply by six:

$$142{,}857 \times 6 = 857{,}142$$

"In this case, the two groups of three numbers have changed places.

"Something quite different happens when we multiply by seven:

$$142{,}857 \times 7 = 999{,}999$$

"Now, let us multiply the number by eight:

$$142{,}857 \times 8 = 1{,}142{,}856$$

"All the numbers appear in the answer except 7. The 7 of our first number is now in two parts, 6 and 1, with 6 on the right and 1 on the left.

"Now, let us multiply our number by nine:

$$142{,}857 \times 9 = 1{,}285{,}713$$

"Look closely at the answer. The only number that is missing is 4. What has happened to it? It seems to have separated into two parts, 1 on the left and 3 on the right.

"We can further show the oddness of the number 142,857 when we multiply by 11, 12, 13, 14, 15, 16, 17, 18, and so on.

"All this has caused the number 142,857 to be considered one of the most mysterious numbers in all of mathematics. I was taught this by the dervish Nô-Elim . . ."

"Nô-Elim?" the prince cried, lighting up. "Did you really know that wisest of men?"

"I knew him very well, My Prince," replied Beremiz. "It was from him that I learned all the principles that I use today in my mathematical studies."

"The great Nô-Elim was a friend of my father's," the prince explained. "Sometime later, after losing a son in a cruel and unjust war, he withdrew from the city and never came back. I made many attempts to locate him but could find no trace of his whereabouts. I came to the conclusion that perhaps he had died in the desert, eaten by panthers. Can you possibly tell me where he is to be found?"

"When I departed from Baghdad," Beremiz replied, "I left him in Khoi, in Persia, with three friends."

"Then, when we return from Mecca, we shall go to the city of Khoi to seek out that great master," the prince replied. "I wish to bring him to my palace. Can you possibly help us in that ambitious task, O Beremiz Samir?"

"My lord," Beremiz replied, "if it is a question of assisting in any way the man who was my guide and master, I shall come with you, if needs be, all the way to India."

And that was how, because of the number 142,857, the matter of our voyage to India, the land of the rajahs, was settled. That number truly is a magical one.

# 24

# Eureka !

*The ill-tempered Tara Tir. The epitaph of Diophantus. Hiero's problem. Beremiz frees himself from a dangerous enemy. A letter from Captain Hassan. The cube roots of 8 and 27. A passion for calculus. The death of Archimedes.*

The threatening presence of Tara Tir was troubling my spirit. The ill-tempered sheik, who had been out of Baghdad for some time, had been seen the preceding night in the company of assassins, prowling in our street. He was no doubt preparing an ambush for the unsuspecting Beremiz, who, busy with his studies, had no sense of the danger that was following him like a dark shadow.

I spoke to him about Tara Tir and reminded him of the careful warning Sheik Iezid had given us.

"Such fear is groundless," he replied, paying no attention to my warning. "I do not believe in these threats. What interests me at the moment is the solution of a problem that lies in the epitaph on the tomb of the famous Greek geometer Diophantus.

"The stone on the grave of Diophantus reveals his age by means of an arithmetical artifice, quite wonderful to contemplate, as follows:

> The gods granted him childhood for a sixth of his life, and a twelfth for his adolescence. A barren marriage took up a seventh of his life. Five years passed, and then a child was born to him. No sooner had this child reached half the age of its father than it died. Diophantus lived for four more years, drowning his pain in the study of numbers, and then gave up his life.

"A study of that text will reveal to you that his age was eighty-four. It is possible that Diophantus was so busy solving arithmetic problems of indeterminacy for so much of his long life that he never thought of finding a solution to the problem by King Hiero, since it does not appear in his works."

"What problem is that?" I asked him. And he told me the following story:

"Hiero, king of Syracuse, delivered to his goldsmiths a quantity of gold to be made into a crown he wished to offer up to Jupiter. When the king received the finished crown, he verified that it weighed the same as the gold he had sent, but the color led him to think that some silver might have been mixed with the gold. He took his doubts to the geometer Archimedes.

"Archimedes, having proved that gold loses fifty-two thousandths of its weight in water, whereas silver loses ninety-nine thousandths, weighed the crown in water and discovered that the difference in its weight indicated the presence of silver mixed with the gold.

"They say that Archimedes took a long time over Hiero's problem. One day in his bath, he hit on the means of solving it and, excited, jumped out and ran to the palace crying, 'Eureka! Eureka!' which means 'I found it! I found it!' "

While we were talking, a visitor appeared, Captain Hassan Maurique, head of the sultan's guard. He was a big man, very easy and obliging. He had heard of the case of the thirty-five camels and, from then on, never ceased to praise the skills of the Man Who Counted. Every Friday on his way back from the mosque, he paid us a visit.

"I never thought," he declared, "that mathematics was so prodigious. I was thrilled by your solution to the problem of the camels."

I took him up to a balcony that overlooked the street, while Beremiz was still busy, and I spoke to him of the danger that threatened us from the hateful Tara Tir.

"There he is," I said, pointing beside the fountain. "Those men with him are dangerous assassins. Given the least chance, they will fall on us. Tara Tir harbors a deep resentment against Beremiz, and I am afraid, since he is a violent and angry man, that he will take revenge. I have seen him spying on us a number of times."

"What are you telling me?" exclaimed Captain Hassan. "I could not have imagined that such a thing might happen. How could a thug like that disturb the life of a wise man like the Man Who Counted? By the Prophet, I'm going to put an end to this matter at once."

When he had gone, I went back to my room and lay down, smoking quietly for a while.

However violent Tara Tir might be, Captain Hassan was no one to trifle with, and he went into action on our behalf. An hour later, I received the following letter from him:

> Everything is resolved. The three assassins have been summarily executed. Tara Tir has been given eight lashes and has paid a fine of twenty-seven pieces of gold. He has been told to leave the city immediately, and I have sent him to Damascus under guard.

I showed the captain's letter to Beremiz. Thanks to this news, we could now live in Baghdad in peace. "Very interesting," remarked Beremiz. "It is really quite curious. This letter makes me remember a strange numerical relationship between the numbers 8 and 27."

I looked surprised at his remark, but he continued, "Excluding the number 1, 8 and 27 are the only numbers equal to the sum of their cubes. Look:

$$8^3 = 512$$
$$27^3 = 19{,}683$$

The sum of the numbers in 512 is 8, and the sum of the numbers in 19,683 is 27."

"You are quite astonishing, my friend!" I exclaimed. "Busy with your cubes, you have forgotten all about being threatened by dangerous assassins."

"Mathematics, O man of Baghdad, takes our attention so much that at times we lose ourselves and forget the dangers that surround us. Do you remember how the great geometer Archimedes died? When the city of Syracuse was taken by storm by the forces of Marcellus, the Roman general, Archimedes was wholly absorbed in a problem whose solution he was working out by drawing in the sand, oblivious to wars and death. The pursuit of truth was all that interested him. A Roman soldier found him and ordered him to come before Marcellus. The sage told him to wait a moment until he had finished the problem he was working on. The soldier insisted and seized him roughly by the arm. 'Be careful! Watch where you're stepping!' cried the sage to the soldier. 'Don't erase that drawing!' Annoyed at not being instantly obeyed, the Roman struck him, putting an end to the life of the wisest man of the day.

"Marcellus, who had given strict orders that Archimedes' life was to be saved, did not hide his sorrow at the death of Archimedes. Over his tomb, he had a stone erected on which was engraved a circle inside a triangle, a figure in memory of one of the theorems of the famous geometer."

Beremiz finished and came over to me, placing a hand on my shoulder. "Don't you think, my friend, that it would be proper to include the wise man of Syracuse among those who are martyrs to geometry?"

What could I say to him?

The tragic end of Archimedes reminded me of the treacherous and envious Tara Tir.

Were we really free of that ill-tempered salt vendor? Might he not later return from his exile in Damascus to cause us more harm?

Standing close to the window with his arms crossed, Beremiz was watching the men coming from and going to the market with a somewhat sad expression on his face. I decided to intrude on his thoughts and take him out of his melancholy.

"What is this?" I asked him. "Are you sad? Do you feel a longing for your own country, or are you simply planning new calculations? Is it mathematics or nostalgia?"

"My friend of Baghdad," he replied, "nostalgia and calculations are not unrelated. One of our most inspired poets has put it as follows:

*Nostalgia can be calculated*
*also using numbers.*
*It is distance multiplied*
*by a factor of love.*

I do not think, however, that nostalgia, once reduced to a formula, can be measured in numbers. When I was a child, I heard my mother many times singing as follows:

*Nostalgia is an old song.*
*Nostalgia is someone's shadow.*
*Only time will carry it off*
*When it carries me off as well.*

# 25

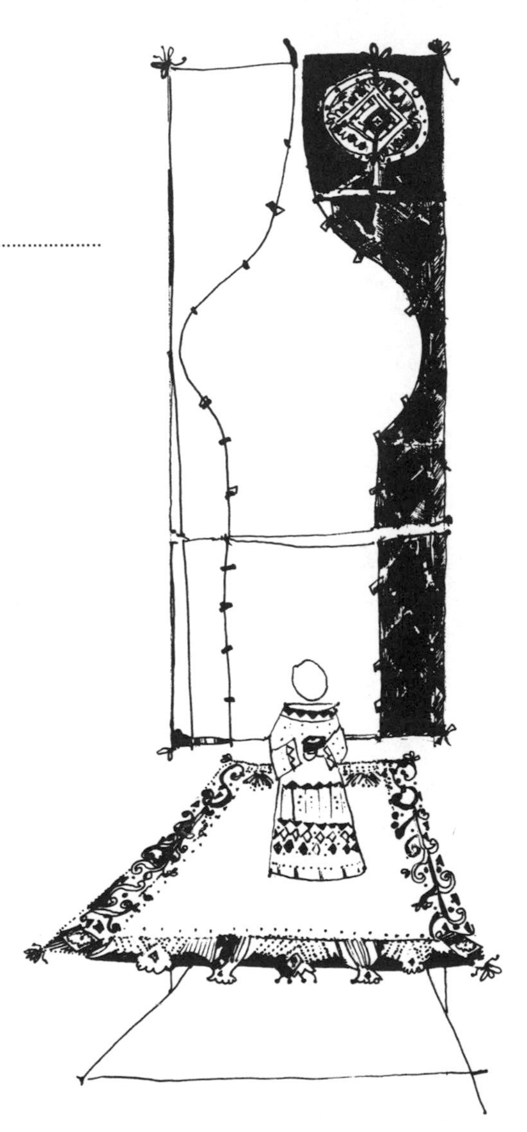

# The questions commence

*Beremiz is summoned to the palace once again. A strange surprise. A difficult match: one against seven. The return of the mysterious ring. Beremiz is presented with a blue carpet. Verses that stir a full heart.*

The first night after Ramadan, arriving at the caliph's palace, we were informed by an old scribe who worked with me that the sovereign was preparing a strange surprise for our friend Beremiz.

A portentous event lay ahead of us. The Man Who Counted was going to have to compete before an audience with seven mathematicians, three of whom had arrived from Cairo the day before. What was to be done? With that challenge ahead, I tried to encourage Beremiz, telling him that he must have absolute confidence in his powers, which had been proved so many times. He reminded me of a proverb of his master Nô-Elim: "He who has no confidence in himself does not deserve the confidence of others."

Filled with a heavy apprehension, we went into the palace.

The huge reception hall was ablaze with light and filled with courtiers and distinguished sheiks. To the right of the caliph sat the young prince Cluzir Shah, guest of honor, accompanied by eight Hindu doctors wearing rich robes of golden velvet and strange turbans from Kashmir. To the left of the throne were seated the viziers, the poets, the judges, and the most distinguished members of Baghdad society. On a dais, on several silk cushions, sat the seven sages who were going to question the Man Who Counted. At a sign from the caliph, Sheik Nuredin Zarur took Beremiz by the arm and led him solemnly to a kind of rostrum set up in the middle of the splendid hall.

The faces of those present were alive with expectation, although not all of them wished success to the Man Who Counted.

A huge black slave sounded a heavy silver gong three times. All the turbans bowed. The strange ceremony was about to begin. My thoughts, I must confess, were in a whirl.

A high priest held the Holy Book and read in a slow and steady voice a prayer from the Koran:

> In the name of Allah, the wise and merciful, Creator of all worlds. We praise thee, O Lord, and beg thy divine aid. Lead us by the strait way, the way of those chosen and blessed by thee.

When the last word had echoed through the galleries of the palace, the king took two steps forward, stopped, and said, "Our friend and ally, Prince Cluzier Shah, lord of Lahore and Delhi, has asked me to propose to the learned men of his retinue the chance to witness the wisdom and skill of the Persian mathematician, secretary to Vizier Ibrahim Maluf. It would be ungracious not to attend to that request on the part of our illustrious guest. And so, seven of the wisest and best-known sages of Islam are going to present Beremiz, the Man Who Counted, with a series of questions related to the science of numbers. If Beremiz can answer these questions, he shall receive, I promise, a reward that will make him one of the most envied men in Baghdad."

At that moment, we saw the poet Iezid approach the caliph. "Lord of all believers!" said the sheik. "I have here something that belongs to Beremiz, a ring that was found in my house by one of the slaves. I wish to return it to him before he undertakes this most important test to which he is to be submitted. It may be that it is some kind of charm, and I do not wish to deprive him of any supernatural help it may give him."

And after a brief pause, the noble Iezid continued, "My well-beloved daughter Telassim, the true treasure among the treasures in my life, asked me to allow her to offer to the Persian mathematician, her master in the science of numbers, this rug on which she has embroidered a border. This rug, if Your Eminence will allow it, is to be placed under the cushion set aside for Beremiz when he is put to the test by the seven most famous sages of Islam."

The caliph ordered the ring and the rug to be taken immediately to the Man Who Counted. Sheik Iezid himself, cordial and friendly as always, handed Beremiz the small box containing the ring. Then, at a signal from the sheik, a young slave appeared carrying a small blue rug that was laid under Beremiz's green cushion.

"All that is a spell, a charm for good luck," said a low voice behind me, the voice of a thin, old man in a blue tunic. "That young Persian knows a lot about magic. That blue rug seems to me somewhat mysterious."

How could they understand, the majority of those present, that Beremiz's aptitude for calculation was the fruit of his pure intelligence? The unlettered, when something is beyond their understanding, always attribute what they do not understand to magical powers. However, the intelligence and culture of the sheiks in attendance was sufficiently high for them to understand that what was taking place was a matter of pure intelligence. Beremiz was to be put to the test by those most skilled in a discipline in which we Arabs have always excelled. Would he manage to meet the test?

Beremiz seemed deeply moved at receiving the ring and the carpet. Even at a distance I could see at that moment that something was affecting him deeply. When he opened the little box, his bright eyes misted over. I discovered later that along with the ring the gentle Telassim had enclosed a slip of paper on which Beremiz read, "Courage. Trust in God. I pray for you." And the carpet, was there really something magical about it, as the old man in the blue tunic had suggested?

It was not magic. That rug, which in the eyes of the sheiks and the wise men was no more than a small gift, contained, written in Kufic characters that Beremiz alone knew how to read and understand, some verses that touched his heart. Those verses, which I later translated, had been embroidered by Telassim to look like arabesques along the borders:

*I love you, my dear one. Forgive me my love.*
*You consoled me, a bird strayed in its flight.*
*When my heart was touched, it lay unveiled and open to the weather.*
*Enfold it in your mercy, my dear one, and forgive me my love.*

*If you cannot love me, my dear one, forgive me my pain.*
*I will return to my song. I will stay seated in the darkness.*
*I will cover with my hands my naked modesty.*

Was Sheik Iezid aware of that double message of love? I do not know quite why that idea came to me so strongly. Only later, as I have said, did Beremiz tell me the secret.

Only Allah knows the truth!

A profound silence fell over the whole sumptuous company. In the rich reception hall of the caliph's palace, it was about to begin, this extraordinary contest that had never before taken place under the skies of Islam.

Allah!

# 26

# One for the book

*Our encounter with a famous theologian. The problem of the life to come. Every Muslim must know the Holy Book. How many words are there in the Koran? How many letters? A deception by Beremiz.*

The wise man appointed to begin the questioning solemnly got to his feet. He was in his eighties and inspired in me a deep respect. Like a prophet, he had a long white beard, which reached to his broad chest. "Who is that fine old man?" I asked in a low voice a doctor with a thin and tanned face who was sitting next to me.

"He is the celebrated sage Mohadeb Ibhage Abner Rama," he answered. "They say that he knows more than fifteen thousand maxims on the Koran. He is a professor of theology and rhetoric."

The sage Mohadeb pronounced his words in a strange manner, syllable by syllable, as if determined to measure the sound of his own voice.

"I am going to question you, O counting man, on a matter of prime importance to a Muslim. Before he studies Euclid or Pythagoras, he of the Islamic faith must have a profound knowledge of his religion, for life is inconceivable if truth and faith are separated. He who does not dwell on the life to come and the salvation of his soul and does not know the precepts of Allah and the commandments does not deserve to be called a wise man. So I want you to give us, without any hesitation, fifteen numerical references from the Koran, the Book of Allah. Among those fifteen examples you must include

1. The number of chapters in the Koran.
2. The exact number of verses.
3. The number of words.
4. The number of letters in the Holy Book.
5. The exact number of prophets mentioned in its pages."

He continued, his voice deepening, "And apart from the five instances I have asked you for, I want you to give us ten more that have to do with numbers. Pray proceed."

A deep silence followed his words, as everyone waited for Beremiz to speak. With remarkable calm, the young counting man replied as follows:

"O wise and venerable Father! The Koran consists of 114 chapters, 70 of them dictated in Mecca and 44 of them in Medina. These are divided into 611 sections and contain 6,236 verses, 7 in the first chapter and 8 in the last. The longest chapter is the second, with 280 verses. In the Koran, there are 46,439 words and 323,670 letters, each one of which contains ten special virtues. Our Holy Book mentions the names of twenty-five prophets. Jesus, the son of Mary, is referred to nineteen times. The names of five animals are taken as epigraphs to 5 chapters: the cow, the bee, the ant, the spider, and the elephant. Chapter 102 is entitled 'The Reply of the Numbers.' That chapter is distinguished by the warning it gives in its 5 verses to those involved in sterile arguments about numbers, which have no importance at all in the spiritual progress of men."

At this point, Beremiz paused briefly and then added, "In answer to your question, these are numerical references in the Book of Allah. In the reply I have just given you, there is one error that I hasten to point out. Instead of fifteen, I have given you sixteen."

"By Allah!" exclaimed the old man in the blue tunic seated behind me. "How can any man know so many numbers, so many things, by heart? It is amazing! He knows even how many letters there are in the Koran!"

"He studies a great deal," muttered his neighbor, a fat man with a scar on his chin. "He studies a great deal, and he remembers everything. I have heard that from a number of people."

"Remembering isn't everything," whispered the old man. "For example, I can't even remember the ages of my cousins."

I was extremely annoyed by all these whisperings around me. But the fact is that Mohadeb confirmed all the references Beremiz had supplied, even the number of letters in the Holy Book.

They told me that Mohadeb, the theologian, was a man who chose to live in poverty, and it must have been true. Allah deprives many wise men of riches, since riches and wisdom seldom occur together.

Beremiz had brilliantly overcome the first challenge offered him in that fearsome debate, but six more were still to come.

"May it be Allah's will," I thought, "that the rest will go as this has gone, and that all will end well."

# 27

# History in the making

*A wise historian questions Beremiz. The geometer who could not see the sky. Greek mathematics. Praise for Eratosthenes.*

The first challenge having been met in detail, the second wise man took up the questioning of Beremiz. He was a famous historian who had given classes for twenty years in Córdoba and who later, for political reasons, moved to Cairo, where he resided under the protection of the caliph. He was a short man with a bronzed face and an elliptical beard. His eyes looked dead, lifeless. Here is what he said to the Man Who Counted:

"In the name of Allah, the Wise and Merciful! They are mistaken, those who think that the worth of a mathematician lies in the skill with which he calculates or applies the banal rules of calculation. In my view, the true mathematician is he who has a thorough knowledge of the progress of mathematics through the centuries. To study the history of mathematics is to pay homage to those geniuses who have elevated and dignified previous civilizations through their intelligence and have been able to reveal some of the most profound mysteries of nature, managing through science to better our wretched human condition. Throughout the pages of history, we honor our glorious ancestors who formed the science of mathematics, and we keep alive the names of the works they left behind them. And so I wish to pose to the Man Who Counted a question on an interesting case in the history of mathematics. What was the name of the famous geometer who committed suicide because he could not see the sky?"

Beremiz thought for a moment and answered, "His name was Eratosthenes, the mathematician of Cyrenaica, who was first educated in Alexandria and later in the School of Athens, where he learned Plato's doctrines. Eratosthenes was appointed to direct the great library in the University of Alexandria, a post he occupied until the end of his days. Besides possessing an enviable depth of scientific and literary knowledge, which put him among the wisest men of his day, Eratosthenes was a poet, an orator, a philosopher, and, even more, an all-around athlete. It is enough to mention that he was a winner of the pentathlon, the fivefold contest in the Olympic Games. Greece then enjoyed a golden age in science and letters. It was the homeland of the epic poets, who declaimed their verses to a musical accompaniment at the banquets and gatherings of kings and great leaders.

"It should be pointed out that, among the most celebrated and cultivated of the Greeks, Eratosthenes was considered an extraordinary man who hurled the javelin, wrote poems, defeated the great runners, and solved astronomical problems. Various works of his have come down to posterity. He presented King Ptolemy III of Egypt with a table on which the prime numbers were etched on a metal plate, with those numbers with multiples marked with a small hole. And so they give the name of Eratosthenes' sieve to the process the wise astronomer used to draw up his table.

"Because of an affliction to his eyes, which he caught on the banks of the Nile in the course of a journey, Eratosthenes went blind. He who had cultivated astronomy with passion was prevented from seeing the sky or admiring the incomparable beauty of the firmament on starry nights. The blue light of Sirius could never penetrate that black cloud that covered his eyes. Overcome by such misfortune, and not able to bear the burden of blindness, the sage committed suicide by allowing himself to die of hunger, shut up in his library."

The wise historian with the dead eyes turned toward the caliph and declared, after a brief silence, "I consider myself well satisfied with the brilliant historical exposition delivered by the Persian calculator. The only famous geometer who committed suicide was indeed Eratosthenes the Greek poet, astronomer, and athlete, close friend of the most famous Archimedes of Syracuse. Allah be praised!"

"By the beauteous fountain of paradise!" cried the caliph enthusiastically. "How many things I have just learned! How many things we are ignorant of! That famous Greek, who studied the stars, wrote poems, and performed athletic feats, is worthy of our deepest admiration. From now on, whenever I look at the sky on starry nights, whenever I see the star Sirius, I shall think of the tragic end of that wise man who wrote the poem of his death surrounded by a treasure of books he was not able to read."

And, touching the shoulder of the prince, he added, "Now let us see if the third wise man will be able to defeat our counting man!"

# 28

# False hopes

*The memorable contest continues. The third wise man questions Beremiz. False induction. Beremiz shows that a false principle may be suggested by true examples.*

The third wise man who was to question Beremiz was the famous astronomer Abul Hassan Ali, from Alcalá, who had come to Baghdad at the caliph's special invitation. He was tall and bony, with a wrinkled face, and he wore on his right wrist a broad gold bracelet that, according to rumor, had engraved on it the twelve signs of the zodiac. After saluting the king and the nobles, the astronomer turned to Beremiz. His deep voice seemed to resound in the hall.

"The two replies that you have just delivered show that you have a good grounding. You speak of Greek science and the close details of our Holy Book with the same ability and mastery. In the science of mathematics, however, the most interesting part is that reasoning that leads to the truth. A collection of facts is as far from forming a body of knowledge as a mirage in the desert is from being a real oasis. Knowledge must observe facts and, from them, deduce its laws. With the help of these laws we can deal with other facts or improve the conditions of life. All that is true. But how do we arrive at the truth? The following question arises:

"Is it possible in mathematics to arrive at a false rule from true facts? I wish to hear your reply, O calculator, and to have it illustrated by a simple example."

Beremiz waited a moment, deep in thought, then he gathered himself and spoke as follows: "Let us suppose that a mathematician out of curiosity wishes to determine the square root of a four-figure number. We know that the square root of a number is another number, which, multiplied by itself, gives an answer equal to the original number. That is a mathematical axiom.

"Let us suppose further that the mathematician, picking three numbers for his experiment, chooses the following: 2,025, 3,025, and 9,801.

"Let us begin with the number 2,025. Making the appropriate calculations, we discover that the square root of that number is 45, which is to say that 45 × 45 equals 2,025. But we can also show that 45 can be obtained by adding 20 and 25, which are parts of the number 2,025, divided in the middle. The same can be demonstrated with the number 3,025, whose square root is 55. The number 55 can also be arrived at by adding 30 and 25, both parts of the number 3,025. The same is true of the number 9,801. Its square root is 99, that is to say 98 plus 1. Given these three examples, a careless mathematician might feel inclined to pronounce the following rule:

"To calculate the square root of a four-figure number, divide the number in the middle into two separate numbers, and add these numbers. The answer will be the square root of the given number.

"That rule, which is obviously erroneous, was arrived at from three real examples. It is impossible in mathematics to arrive at the truth by simple observation, but even so, it is important to take special care to avoid such false induction."

The astronomer Abul Hassan, obviously satisfied with Beremiz's reply, declared that he had never heard such a simple and interesting explanation of false induction in mathematics.

At a signal from the caliph, the fourth wise man took his turn and got ready to ask his question. His name was Jabal ibn-Wafrid, a poet, a philosopher, and an astrologer. In his native Toledo, he had become well known as a teller of stories. I shall never forget his venerable presence or his serene and kindly gaze. He moved to the edge of the dais and, addressing the Man Who Counted, began, "In order for you to understand my question, I must first tell you an old Persian legend."

"Tell it then, O eloquent sage!" said the caliph. "We are eager to hear your wise words, which are as drops of gold to us who listen."

The wise man of Toledo, in a voice as firm and steady as the progress of a caravan, told the following story.

# 29

# Single-handed success

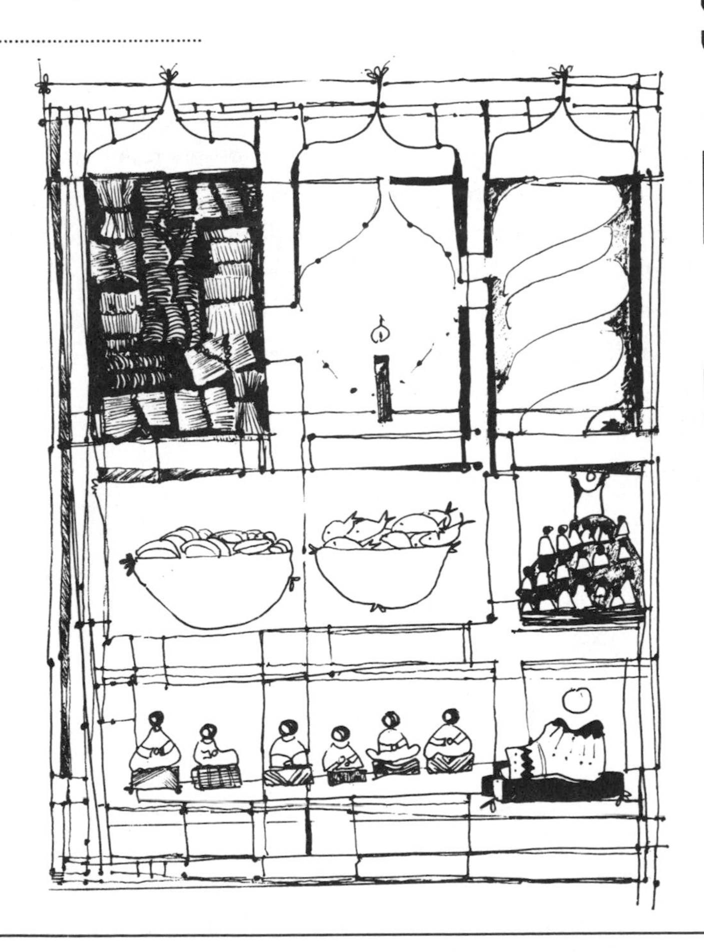

*We hear an old Persian legend. The material and the spiritual. Problems human and superhuman. The most famous multiplication of all. The Sultan fiercely chastises intolerance among the sheiks of Islam.*

A powerful king who governed Persia and the great plains of Iran heard a certain Dervish once declare that the true wise man must know and distinguish the spiritual and the material in life. That king was called Astor and was known familiarly as the Serene.

One day he summoned the three wisest men in all of Persia, gave to each of them two silver dinars, and spoke to them as follows: "In this palace there are three identical rooms, all of them completely empty. Each one of you is commissioned to fill up one of these rooms, but in performing this task you must not spend any more than you have just received."

The problem seemed a truly difficult one. Each wise man had to fill up an empty room, spending no more than the insignificant sum of two dinars. The wise men set out to fulfill the difficult commission set them by King Astor.

Some time later, they returned to the throne room. The king, eager to hear their solutions to the problem, questioned them in turn.

The first spoke as follows: "My lord, I have spent the two dinars, and the room is now completely full. My solution was a practical one. I bought a number of sacks of hay, and with them I filled the whole room from floor to ceiling."

"Very good!" exclaimed King Astor. "Your solution was truly imaginative. In my opinion, you are well aware of the material part of life, and from that vantage point you are well able to take on the problems that life presents to you."

The second wise man, after bowing to the king, gave his answer as follows: "In carrying out my task, I spent only half a dinar. Let me explain. I bought a candle and lit it in the empty room. Now, O King, you may see for yourself. The room is completely filled, filled with light."

"Bravo!" exclaimed the king. "Your solution is quite brilliant. Light represents the spiritual part of life. Your spirit, it seems to me, is well able to face the problems of existence from the spiritual point of view."

Then the third wise man spoke as follows: "O King of the four corners of the world! At first I thought of leaving the room exactly as I found it. It would have been easy to say that the room was not empty since, obviously, it was filled with air and darkness. I did not, however, wish to seem guilty of laziness and trickery, so I decided to act, as did my companions. I took a handful of hay from the first room, set it on fire from the candle, and extinguished it, so that the room filled up entirely with smoke. As you might suppose, this cost me nothing. The sum you gave me is intact, and the room is full, full of smoke."

"Admirable!" exclaimed Astor the Serene. "You are the wisest of men in Persia, and perhaps in the whole world, since you know how to combine the material with the spiritual in achieving perfection."

The sage of Toledo finished his story and, turning toward Beremiz, addressed him in a friendly manner: "My wish is, O calculator, that you prove, as did the third sage in my story, that you are able so to unite the material and the spiritual, and to solve not just human problems but problems of the spirit. My question then is as follows: Which is the famous act of multiplication, which all histories mention and all men of culture know well, which uses only one factor?"

That question took the illustrious gathering by surprise. Some of those present did not conceal their impatience. A judge at my side grumbled in irritation, "That question is outrageous!"

Beremiz, after a moment's thought, replied as follows: "The only multiplication using a single factor, known to all historians and men of culture, is the multiplication of loaves and fishes performed by Jesus, the son of Mary. In that multiplication there is only one factor: the miraculous power of the will of God."

"An excellent reply!" said the sage of Toledo. "It is the best answer I have ever heard, and the Man Who Counted has solved my problem irrefutably. Praise be to Allah!"

Some of the more intolerant among the faithful looked at one another in astonishment, and there was much whispering. The caliph interrupted in a loud voice: "Silence, all of you! It is for us to venerate Jesus, the son of Mary, whose name is mentioned nineteen times in the Holy Book of Allah."

He turned to face the fifth wise man and said in a friendly voice, "We are waiting for your question now, Sheik Nascif Rahal. You are next."

At the king's command, the fifth wise man got to his feet. He was a short fat man with white hair. Instead of a turban, he wore a small green cap. He was well known in Baghdad since he gave classes in the mosque and elucidated to scholars obscure points in the sayings of the Prophet. I had seen him two or three times coming out of the bathhouse. He spoke nervously, in a somewhat aggressive manner.

"The value of any sage can be measured only by the power of his imagination. Numbers selected by chance, historical deeds remembered in detail, have only a momentary interest; after a time they are forgotten. How many of you remember the number of letters in the Koran? There are numbers, names, words, even whole books that are doomed to be forgotten irretrievably. Knowledge alone does not make a man wise. And so I am going to test the worth of this Persian calculator who appears before us by asking him a question that cannot be answered simply by memory or ability alone. I would like Beremiz Samir to tell us a story, a simple fable, in which there should appear a division of three by three that is suggested but not carried out, and another division of three by two that is carried out but leaves no remainder."

"A fine idea!" whispered the old man in the blue tunic. "We are going to get away from calculations that nobody understands and hear a story instead."

"But that story will have numbers, I'm sure of it," grumbled the doctor beside me in a low voice. "You will see, my friend. Everything comes down to calculations, numbers, and problems."

"I hope not," said the old man.

I was somewhat startled and taken aback by the demands made by the fifth wise man. How could Beremiz invent in a moment a story that encompassed a division suggested but not realized or, even more difficult, a division of three by two leaving no remainder, when logic decrees that three divided by two must leave a remainder of one? But I put my anxieties to one side and trusted in the imagination of my friend and in the goodwill of Allah.

The Man Who Counted, after running through his memory for a few moments, began to tell the following story.

# 30

# Three of a kind

*The Man Who Counted tells a story. The tiger suggests how to divide three by three. The jackal proposes the division of three by two. How to arrive at a quotient in the mathematics of the strong. The sheik in the green cap praises Beremiz.*

In the name of Allah, the Wise and Merciful!

On one occasion, the lion, the tiger, and the jackal left the dark cave where they lived and set out on a friendly pilgrimage to wander through the world in search of some region rich in flocks of tender sheep.

In the middle of the great jungle, the fearsome lion, who naturally led the group, felt a weariness in his rear paws and, throwing back his huge head, let loose a roar so fierce that the nearest trees trembled.

Startled, the tiger and the jackal looked at each other. That terrifying roar with which their dangerous leader disturbed the silence of the forest meant to the other two animals, loosely translated, "I am hungry!"

"I can understand your impatience," said the jackal anxiously to the lion. "But I can assure you that in this jungle there is a secret path that no one knows, and it will lead us quickly to a small village, almost a ruin, where there is abundant hunting in easy reach of our claws . . ."

"Then let us go, jackal!" roared the lion. "Lead me to this wonderful place!"

At evening, led by the jackal, the travelers reached the top of a low mountain, from where they could see a broad, green plain. In the middle of the plain, they saw three gentle animals grazing, unaware of any danger: a sheep, a pig, and a rabbit.

At the sight of such easy and sure prey, the lion shook his abundant mane with obvious satisfaction and, his eyes shining with greed, turned to the tiger and said, in a seemingly friendly tone, "Now, my fine tiger! I see there three splendid and tasty mouthfuls: a sheep, a pig, and a rabbit. You, who are an expert, are charged with dividing them among the three of us. Do it justly and equably—divide these three animals among three hunters in a brotherly fashion."

Flattered at the invitation, the vain tiger, after protesting his inadequacy with roars of false modesty, answered as follows: "O King, the division you have generously proposed is quite simple and can be done with relative ease. The sheep, which is the choicest and tastiest, could satisfy the hunger of a whole pride of desert lions, and so it is yours alone, yours absolutely. And that pig—skinny, dirty, and sad, not equal to one leg of that fat sheep—will be for me, since I am modest and content with little. And finally, that small and miserable rabbit with little flesh, unworthy of the palate of a king, will fall to our friend the jackal as a reward for the valuable directions he gave us."

"Idiot! Egoist!" roared the lion, terrifying in his fury. "Who taught you to make divisions like that? Imbecile! Who has ever seen three divided by three giving a result like that?"

And raising his paw, he swiped the head of the unsuspecting tiger so fiercely that he fell dead a few feet away. Then, turning to the jackal, who had witnessed that tragic division of three by three with some horror, the lion spoke as follows: "Now, my dear jackal, I have always had the highest regard for your intelligence. I know you are a most ingenious and clever animal, and I know no other who can resolve the most difficult problems as cleverly. And so I charge you to make this division, such a simple and trivial one, which the stupid tiger, as you have just seen, could not solve satisfactorily. Take a look, friend jackal, at these three mouth-watering animals: the sheep, the pig, and the rabbit. There are two of us and three meals to divide. So do your division—I wish to know which is my exact share."

"I am no more than your humble and poor servant," sniveled the jackal meekly. "I must blindly obey my orders. Like a wise calculator, I am going to divide these three animals by two, a simple division! The most mathematically sure and fair division is the following: that splendid sheep, a feed worthy of a king, is for your royal mouth alone, since you are unquestionably the king of all animals. That appetizing pig, whose gentle grunts you can hear from here, must also be destined for your royal palate, since those who know say that the flesh of a pig gives both strength and energy to lions. And the skittish rabbit with its large ears is also for you as a savory bite, since kings traditionally, at the finest banquets, must always enjoy the most delicate fare."

"Incomparable jackal!" exclaimed the lion, enchanted with the division he had just heard. "How wise and harmonious your words always are! Who taught you that wonderful trick of dividing three by two so surely and perfectly?"

"It was the justice you just dealt out a moment ago to the tiger for not knowing how to divide three by two when one of the two is a lion and the other a mere jackal. In the mathematics of the strong, I always say, the quotient is always clear, while to the weak must fall only the remainder."

And from that day on, suggesting such divisions, inspired by meanness, the ambitious jackal decided that he could live in tranquillity only as a parasite, receiving only the leftovers from the lion's feasts.

But he was wrong.

After two or three weeks, the lion, angry and hungry, tired of the jackal's servility, ended up killing him as he had the tiger.

And the moral is that the truth must always be told, a thousand and one times, for the punishment of God is closer to the sinner than are his own eyelids.

"And here, O most wise judge!" said Beremiz in conclusion, "is the simplest of fables in which occur two divisions. The first was a division of three by three, suggested but not realized, the second, of three by two, realized with no remainder."

A deep silence followed the words of the Man Who Counted. All those present awaited with lively interest the verdict of the grave wise man.

Sheik Nascif Rahal, after adjusting his green cap nervously and stroking his beard, gave his judgment with seeming reservation: "The story you have told complied perfectly with my demands. I confess that it was new to me, and, in my view, it is a worthy tale. Aesop the Greek himself could not have done it better. That is my opinion."

Beremiz's story, approved by the sheik in the green cap, pleased all the viziers and noblemen present. Prince Cluzir Shah, the king's guest, spoke in a loud voice to the whole company:

"The story we have just heard contains a moral lesson. Wretched adulators who fawn in the courts, who crawl on the carpets of the powerful, may in the beginning gain something by their servility, but in the end they are always punished, since God's punishment is always close at hand. I shall tell it to all my friends and acquaintances when I return to my own country."

The caliph, too, thought Beremiz's story quite wonderful and said also that that notable division of three by three ought to be preserved in his archives, since the story, because of its moral conclusion, deserved to be written in gold letters on the transparent wings of the white butterfly of the Caucasus.

Immediately, the sixth wise man stepped forward.

He was from Córdoba, in Spain, where he had lived for fifteen years before having to flee for having incurred the wrath of his sovereign. He was middle-aged, with a round face and an open, humorous expression. His admirers said that he was most adept at writing humorous and satirical poems directed against tyrants. For six years he had worked in Yemen as a simple guide.

"Emir of the world!" he began, addressing the caliph. "I have just heard with true satisfaction the splendid fable about the division of three by two. That story contains, in my view, a great lesson and a profound truth, a truth as clear as the sun at noon. I must confess that moral precepts come alive when they are presented in the form of stories or fables. I know a story that has no divisions, square roots, or fractions, but that contains a problem in logic that can be solved only by means of pure mathematical reasoning. I shall tell it in the form of a story, and we shall see how our excellent calculator shall solve the problem contained in it."

And the wise man of Córdoba told the following story.

# 31

# In black and white

*The wise man of Córdoba tells a story. The three suitors of Dahize. The puzzle of the five disks. How Beremiz explained the reasoning of a clever suitor.*

Macudo, the famous Arab historian, in the twenty-two volumes of his writings, speaks of the seven seas, the great rivers, famous elephants, the stars, the mountains, the different kings of China, and a thousand other things, but he does not even mention the name of Dahize, the only daughter of King Cassim the Uncertain. It does not matter. In spite of that, Dahize will never be forgotten, since in Arab manuscripts her name appears in more than 400,000 verses in which hundreds of poets rhapsodize over her beauty. The ink used in describing her eyes alone would, if it were turned into oil, be enough to light up the city of Cairo for half a century. You may think I am exaggerating. But, my brothers, I am not, since exaggeration is a form of lying. But let me get to my story.

"When Princess Dahize was eighteen years and twenty-seven days old, her hand in marriage was sought by three princes whose names have passed into legend: Aradin, Benefir, and Comozan.

"King Cassim was uncertain. Of the three rich suitors, how could he choose the one who should marry his daughter? If he were to do so, it could have the following fatal result: he, the king, would gain a son-in-law, but the two unsuccessful suitors would become his bitter enemies. It was a hard decision for a sensitive and cautious king who only wanted to live in peace with his people and his neighbors. He asked Princess Dahize, but she declared only that she would marry the one who was most intelligent.

"Her decision pleased King Cassim, for he saw a simple solution to what seemed an impossible choice. He summoned five of the wisest men in his court and told them to put the three princes through a rigorous test to see which of the three was the most intelligent.

"When they had done so, the wise men reported to the king that all three princes were indeed most intelligent. They were well versed in mathematics, literature, astronomy, and physics. They could solve difficult chess problems, the subtleties of geometry, and all kinds of complex enigmas. 'We do not see any way,' said the wise men, 'of making a clear decision in favor of one of them.'

"After this distressing failure, the king decided to consult a dervish who had a reputation for knowing much about magic and the occult.

"The dervish addressed himself to the king. 'I know only one way that will allow us to decide which prince is the most intelligent of the three—the test of the five disks.'

" 'Then let us do it!' exclaimed the king.

"The three princes were summoned to the palace, and the dervish, showing them five simple wooden disks, said to them, 'Here are five disks, two of them black and three of them white. They are all the same size and weight and are different only in color.'

"Next, a page carefully bound the eyes of the three princes so that they could see nothing. The old dervish then picked three disks at random and fastened one each to the backs of the three suitors, saying as he did so, 'Each one of you has on his back a disk whose color you do not know. You are to be questioned in turn. The one who discovers the color of the disk he is wearing will be declared the winner and will receive the hand of the beautiful Dahize in marriage. The first one questioned can look at the disks of the other two. The second can see only the disk of the third, and the third must make his reply seeing none of the others. The one who gives the correct answer must, in order to prove that he was not simply guessing, justify his answer by clear reasoning. Now, who wants to go first?'

" 'Let me be first,' said Prince Comozan promptly.

"The page removed the bandage from his eyes, and Prince Comozan saw the disks on the backs of his two rivals. The dervish took him aside to hear his answer, but it was wrong. Declaring himself beaten, he withdrew. He had seen the two disks on the backs of the other princes and still not been able to determine the color of his own disk.

" 'Prince Comozan has failed,' said the king in a loud voice, to inform the other two.

" 'Then let me be next,' said Prince Benefir. Once his eyes were uncovered, the second prince saw the disk worn by the third on his back. He motioned to the dervish and whispered his reply to him. The dervish shook his head. The second prince was also mistaken and was given leave to withdraw immediately. Only one was left, Prince Aradin.

"When the king announced that the second suitor had also failed, he approached with his eyes still bandaged and announced in a loud voice the correct color of the disk on his back."

When the story was finished, the wise man of Córdoba turned to Beremiz and said, "In making his answer, Prince Aradin reasoned in such a way as to reach with complete certainty the solution to the problem of the five disks and to win the hand of the beautiful Dahize. Now, I wish you to tell me, first, what his reply was and, second, how he could be so sure of the color of his own disk."

Beremiz, his head down, thought for some time; then, looking up, he gave the following explanation in a firm, clear voice.

"Prince Aradin, the hero of your strange story, said to King Cassim, 'My disk is white,' and in so saying he knew that his answer was true. What, then, was the reasoning that led him to this conclusion? He considered what the first two suitors must have seen.

"The first prince, Comozan, saw the two disks of his rivals but, nevertheless, gave a wrong answer. Why did he err? He erred because he was uncertain. But if he had seen two black disks, he could not have made a mistake or felt any doubt and would have said to the king, 'I see that my rivals each wears a black disk, and since there are only two black disks, mine must be white.'

"But Comozan's answer was wrong, so consequently the disks he saw were not both black. If they were not both black, two possibilities remain. Either they were both white, or one was white and the other black. If Comozan saw two white disks, Aradin reasoned, then the disk on my back must be white. But if Comozan saw one white disk and one black, which of us wore the black disk? If it were I, Aradin reasoned, Benefir would have known the answer.

"In effect, Benefir would have reasoned as follow: I see that the third prince is wearing a black disk. If mine were also black, then Comozan, the first prince, seeing two black disks, could not have made a mistake; but since he was wrong, my disk must be white. But the second prince was also wrong, and so he must also have been uncertain. His uncertainty must have come, Aradin reasoned, from seeing on my back not a black disk but a white one. So Aradin concluded as follows: according to the second hypothesis, my disk then must be white.

"That was how Aradin reasoned," stated Beremiz, "in resolving the problem with absolute certainty and being able to say, 'My disk is white.' "

The wise man of Córdoba then addressed the caliph, declaring that Beremiz's solution to the problem of the five disks was correct and quite brilliant. His simple and clear reasoning was impeccable, and the wise man said he was sure that all those present had understood the problem and would be able to recount it afterwards at any caravan halt in the desert.

A Yemenite sheik, seated in front of me on a red cushion, a dark, evil-faced man, wearing many jewels, murmured to a friend at his side, "Do you hear that, Captain Sayeg? That man from Córdoba says that we have all understood that story of the white disks and the black disks. I doubt it very much. As for me, I didn't understand a single word of it. Only a mad dervish would have thought of putting black and white disks on the backs of the three suitors, don't you think? Wouldn't it have been more practical to hold a camel race in the desert? There would have been a clear winner, and the matter would be perfectly resolved, no?"

Captain Sayeg did not reply, and seemed to ignore the stupid Yemenite who wanted to solve a problem of love by a camel race in the desert.

The caliph declared affably that Beremiz had come through the sixth proof of the contest.

Would our friend, the Man Who Counted, be successful in the seventh and last round? Would he surmount it with the same brilliance? Only Allah knows the truth!

Finally, everything seemed to be going as we had wished.

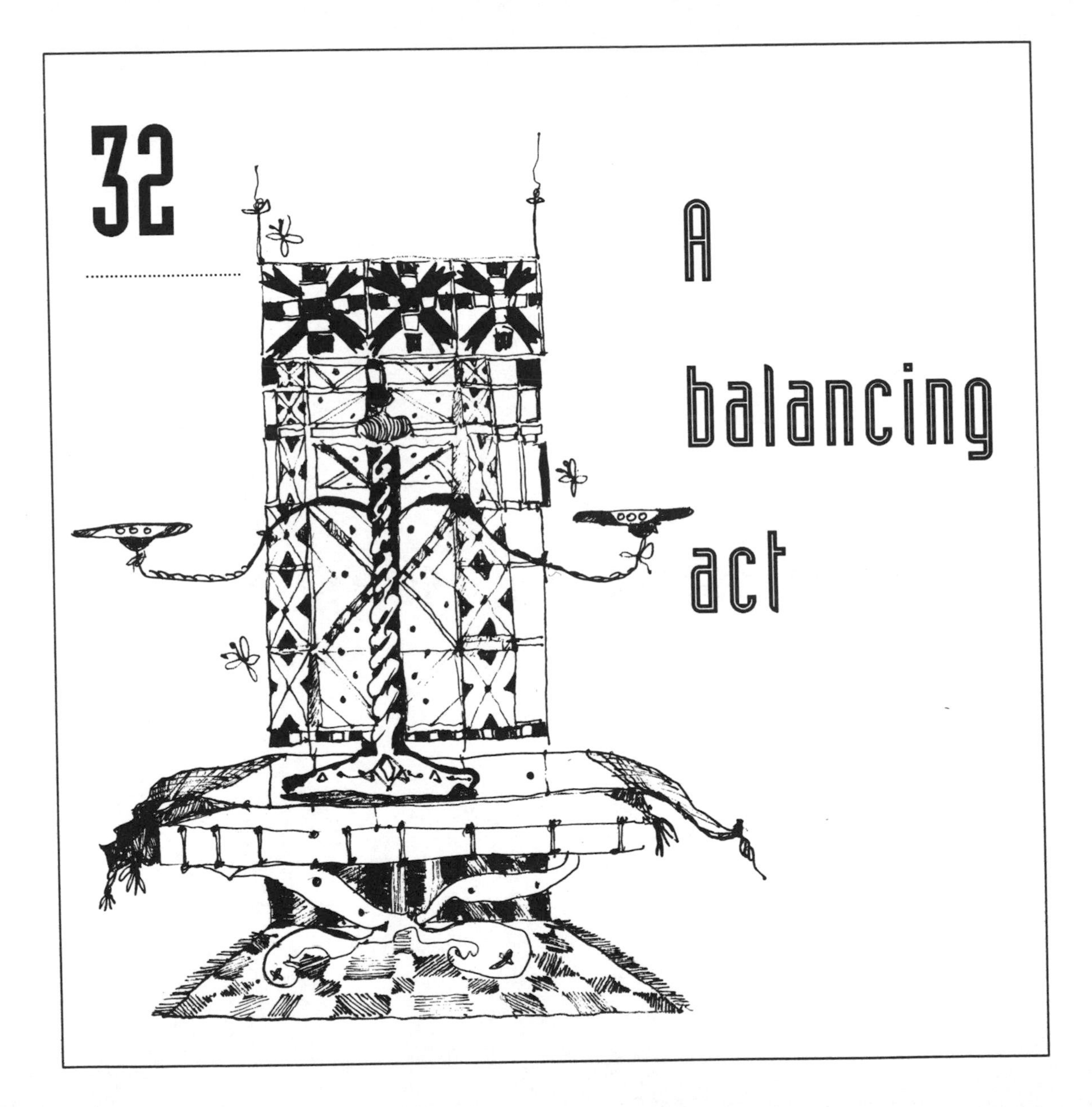
32
A balancing act

*Beremiz is questioned by a Lebanese astronomer. The problem of the lightest pearl. The astronomer quotes a poem in praise of Beremiz.*

His name was Mohildin Ihaia Banabixacar, geometer, astronomer, and one of the most notable names in all of Islam, the seventh and last sage to face Beremiz. He had been born in Lebanon, and his name was inscribed on five mosques and his books were read even by Christians. It would be impossible to find under the skies of Islam anyone of more lucid intelligence or more substantial knowledge.

The learned Banabixacar, his voice clear and precise, spoke as follows: "I am truly delighted with everything I have heard up until now. The distinguished Persian mathematician has just demonstrated over and over again his unquestionable talent. I should like, as my part in this brilliant contest, to present him with an interesting problem I learned when I was young from a Buddhist priest well versed in the science of numbers."

The caliph, showing a lively interest, declared, "Let us hear it then, Arab brother!

We shall listen to your problem with the greatest of pleasure. I hope that our young Persian, who until now has shown himself insuperable in the realms of calculation, will know how to solve the puzzle of the old Buddhist."

The Lebanese sage, seeing that his words had caught the attention of the king and all those present, spoke as follows, looking keenly at the Man Who Counted: "My problem should properly be called 'The Problem of the Lightest Pearl.' "

He went on, "A merchant of Benares, in India, had in his possession eight pearls identical in shape, size, and color. Of these eight pearls, seven were the same weight, while the eighth weighed slightly less than the others. How could the merchant discover which pearl was lighter, using a scale but making only two weighings and not using any weights? That is the problem, and may Allah lead you, O counting man, to a simple and perfect solution."

Hearing the problem so stated, a white-haired sheik wearing a gold collar, sitting by Captain Sayeg, murmured in a low voice, "What an elegant problem! This Lebanese is a man of genius. Praise be to Lebanon, the country of cedars!"

Beremiz Samir, after his customary pause for thought, spoke in a slow and firm voice: "It does not seem to me too difficult, the problem of the Buddhist. A path of clear reasoning can surely lead us to a solution.

"Let us see. There are eight identical pearls, identical in shape, color, size, and brilliance. We have been assured that one of those eight pearls is lighter than the others, while the other seven all weigh the same. The only way of discovering which pearl weighs less than the others is to use a scale, a finely calibrated scale with long arms and light plates. One thing more: the scale must be precise. If I took the pearls two by two and placed them on the scale, one in each pan, it would naturally be easy to discover eventually the light pearl. But supposing it were one of the last pair, I would have been obliged to make four weighings, and the problem demands that I discover the light pearl in only two. The simplest solution seems to me to be the following;

"Let us divide the pearls in three groups: A, B, and C. Group A will consist of three pearls, group B also of three, and group C of the remaining two. In only two weighings, I shall find out which is the light pearl, given that the other seven weigh exactly the same.

"Let us put the groups A and B on the scale, one in each pan. Two things can happen: either groups A and B weigh the same, or one weighs more than the other.

"In the first case, where A and B weigh the same, we can be sure that the light pearl belongs to neither of the two groups, therefore it must be one of the two pearls that make up group C. We shall then weigh these two pearls, one in each pan, and the scale, at this second weighing, will indicate which one is lighter.

"In the second case, where group A weighs less than group B or vice versa, it will be clear that the light pearl is in one of the groups. From that group let us take any two pearls and put them in the pan of the scale for a second weighing. If the scale balances equally, then the third pearl from the group, the one we have put aside, is the light pear, If, however, the balance is unequal, then the light pearl must be in the pan that rises.

"And so, the problem posed by the illustrious Buddhist priest is solved, and I present my solution to our distinguished guest, the sage from Lebanon.

Banabixacar, the astronomer, agreed that the solution offered by Beremiz was indeed impeccable, and he spoke in the following manner: "Only a true mathematician could produce such perfect reasoning. The solution I have just listened to is a true poem in its beauty and simplicity."

And, in homage to the Man Who Counted, the Lebanese astronomer quoted the following verses from Omar Khayyám, a Persian poet of the highest quality and also a celebrated mathematician:

*If you have kept one rose of love close to your heart . . .*
*If you have addressed your humble prayer to one just and supreme God . . .*
*If, raising your cup, you one day sing your praise of life . . .*
*You have not lived in vain . . .*

Much moved, Beremiz bowed his thanks for this homage and placed his right hand on his heart.

# 33

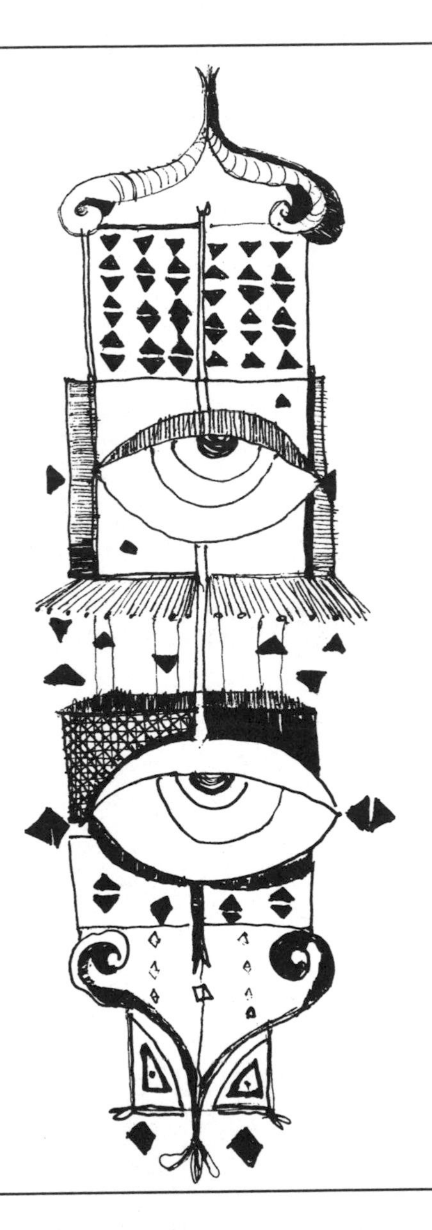

# Eye to eye

*The offer that Caliph al-Mutasim made to the Man Who Counted. Beremiz will not accept either gold, goods, or palaces. He asks instead for a hand. The problem of the black eyes and the blue eyes. Beremiz discovers by reasoning the color of the eyes of the five slaves.*

When Beremiz had finished solving the problem set him by the wise man of Lebanon, the sultan, after conferring in a low voice with two of his counselors, addressed him.

"By the answers you have given to all the questions, you have shown yourself most deserving of the reward I promised you. And so I give you a choice: Do you wish to receive twenty thousand gold dinars, or would you rather have a palace in Baghdad? Would you like to be governor of a province, or would you prefer the post of vizier in my court?"

"O generous King!" replied Beremiz, deeply moved. "I seek neither riches nor titles nor honors nor gifts, since I know that such things are worth nothing. I am not drawn by the prestige that such things would give me, since my spirit does not seek the ephemeral glories of worldly possessions. If, however, you wish to make me the envy of all Muslims, as you previously said, my request is as follows:

I wish to marry young Telassim, daughter of Sheik Iezid Abul Hamid."

This surprising request by the Man Who Counted caused an indescribable uproar. From all the comments that I heard around me, I realized that all those present were convinced that Beremiz was completely mad.

"That man is crazy," murmured the thin old man in the blue tunic behind me. "Crazy, I say. He rejects wealth, he turns his back on fame, and all to marry a young girl he has never seen!"

"He is indeed delirious," added the man with the scar. "Yes, delirious. He asked for the hand of a girl who may very well detest him. By Allah!"

"Could it be the spell of the blue rug?" asked Captain Sayeg in a low voice, not without malice. "Has the blue rug bewitched him?"

"Blue rug, indeed!" exclaimed the old man. "There is no spell for winning the heart of a woman."

I listened to all those remarks while pretending to be thinking of something else. At Beremiz's request, the caliph frowned and looked grave. He summoned Sheik Iezid, and the two of them whispered together for a few moments. What would be the result of their consultations? Would the sheik agree to the betrothal of his daughter?

After a time, the caliph spoke as follows, in the midst of a deep silence: "I shall not oppose your happy marriage with the beautiful Telassim, O Beremiz. My esteemed friend here, Sheik Iezid, whom I have just consulted, will accept you as a son-in-law. I realize that you are a man of character, well educated, and deeply religious. Now, it is true that the beautiful Telassim was promised to a sheik of

Damascus who is now fighting in Spain. But if she herself wishes to alter the course of her life, I will not stand in her way. It is written! The arrow, once in flight, cries happily, "I am free, I am free!" but in truth it deceives itself, for its destiny has been appointed by the aim of the marksman. Such is the case with the young Flower of Islam! Telassim rejects a rich and noble sheik who tomorrow could be vizier or governor, and accepts for her husband a simple Persian calculator. It is written! May it be what Allah wishes!"

The caliph paused and then continued forcefully, "Nevertheless, I lay down one condition. Before those gathered here, you must solve a strange problem set by a dervish from Cairo. If you solve it correctly, you may marry Telassim. If not, you must give up forever this wild dream and you will receive nothing from me. Do you accept the conditions?"

"O Master of all believers!" replied Beremiz firmly. "Only tell me the problem for me to solve it . . ."

And so the caliph spoke: "The problem, put very simply, is the following: I own five beautiful slave girls, recently purchased from a Mongol prince. Two of those young enchantresses have black eyes; the other three, blue eyes. The two with black eyes always give a truthful answer to any question, whereas the three with blue eyes are born liars and never answer with the truth. In a few moments, the five of them will be brought here, all of their faces covered by a heavy veil, which will make it impossible for you to see their faces. You must discover, with no room for error, which of them have black eyes and which blue eyes. You may question three of the five slaves, one question to each one. From the three answers, you must solve the problem and explain the precise reasoning that led you to your answer. Your questions should be quite simple ones, well within the compass of these slaves to answer."

Some moments later, watched curiously by everyone present, the five slave girls appeared in the reception hall, their faces covered with black veils, like phantoms of the desert.

"Here they are," said the emir somewhat proudly. "Two of them as I told you have black eyes and speak only the truth. The other three have blue eyes and always lie."

"What a scandal!" muttered the thin old man. "Imagine my bad luck! My uncle's daughter has black eyes, very black—yet she lies all day long!"

His remark seemed to me out of order. It was a very serious moment, no time for jokes. Luckily, no one paid any attention to the nasty words of the impertinent old man. Beremiz realized that he had reached a decisive moment, perhaps the high point of his whole life. The problem that the caliph of Baghdad had presented him with was original and difficult and could be full of pitfalls. He was free to question three of the slave girls. But how would their replies indicate the color of their eyes? Which of them should he question? How would he know the eye colors of the two he could not question?

There was only one certainty, that the two with black eyes always spoke the truth and the other three invariably lied. But would that be enough? When Beremiz questioned them, the question had to be quite natural, well within the reach of the girl questioned. But how could he be sure of her reply, whether it was true or false? It was really very difficult indeed.

The five veiled slave girls lined up in the middle of the sumptuous hall in complete silence. The sheiks and viziers waited for the solution to this singular problem set by their king with a lively interest. The Man Who Counted approached the first slave girl on the right of the row, at the end, and asked her quietly, "What color are your eyes?"

The girl replied in a language that was apparently Chinese, a language that nobody present could understand. It made no sense to me. Hearing her, the caliph ordered that the other replies be in Arabic, simple and precise.

This unexpected setback made things more difficult for Beremiz. He had only two questions left, and the answer to his first question had been completely lost.

This did not seem to upset him, however, as he approached the second slave and asked her, "What was the reply that your companion just gave?"

The second slave answered, "She said, 'My eyes are blue.' "

That reply cleared up nothing. Had the second slave told the truth, or was she lying? And the first? Was that her real reply?

The third slave girl, in the center of the row, was questioned next by Beremiz.

"What color are the eyes of those two girls I have just questioned?"

And the third girl, the last to be questioned, replied as follows: "The first girl has black eyes and the second blue eyes."

Beremiz paused a moment and then calmly approached the throne, speaking as follows: "Lord of all believers, Shadow of Allah on earth, I have a solution to your problem, arrived at through strict logic. The first slave girl on the right has black eyes. The second has blue eyes. The third has black eyes, and the other two have blue eyes."

At this the five slave girls lifted their veils, uncovering smiling faces. Great sighs arose throughout the reception hall. Beremiz in his faultless intelligence had exactly determined the color of all their eyes.

"All praise to the Prophet!" cried the king. "This problem has been set to hundreds of wise men, poets, and scribes, and at last this modest Persian is the only one who has been able to solve it. How did you come by your answer? Show us how you came to be so certain of your solution."

The Man Who Counted gave the following explanation:

"When I asked the first question—what is the color of your eyes—I knew that the slave girl's answer had to be 'My eyes are black,' because if she had black eyes she would have to tell the truth, or if she had blue eyes, she would have to lie. So there could be only one reply, namely, 'My eyes are black.' I expected that answer; but when the girl replied to me in an unknown language, she helped me enormously. Claiming not to have understood, I asked the second slave girl, 'What was the reply your companion just gave?' and received the second answer—'She said: my eyes are blue.' That reply proved to me that the second girl was lying, since, as I have already shown, it could not have been the reply of the first girl. Consequently, if the second slave was lying, she had blue eyes. This was an important point, O King, in solving the problem. Of the five slave girls, there was at least one whom I had identified with mathematical certainty, namely, the second. She had lied, and so she had blue eyes.

"My third and last question I asked of the girl in the center of the row. 'What color are the eyes of those two girls I have just questioned?' She gave me the following reply: 'The first girl has black eyes and the second blue eyes.' Since I knew already that the second did have blue eyes, what conclusion was I to come to over the reply of the third girl? Very simple. The third girl was not lying, since she was confirming what I already knew, namely, that the second girl had blue eyes. Her reply also told me that the first slave girl had black eyes. Since the third girl was not lying, her words spoke the truth, and therefore she too had black eyes. From there, it was simple to deduce that the two other girls, by exclusion, had blue eyes.

"I can assure you, O King, that in this problem, although no equations or algebraic symbols appear, the perfect solution can be reached only by the rigorous logic of pure mathematics."

So was resolved the caliph's problem. But for Beremiz there soon awaited another, more difficult problem, that of Telassim, the treasure he had dreamed of in Baghdad.

All praise to Allah, who created woman, love, and mathematics!

# 34

# Of life and love

*I bring to a close the story of Beremiz, the Man Who Counted.*

. . . . . . . . . . . . . . . . . . . . . . . . . . . . . . . . . . . . . . . . . . . . . . . . . . . . . . . . . . . . . . . . . . . . . . .

In the third moon of Rhegeb, in the year 1258, hordes of Tartars and Mongols, under the command of a grandson of Genghis Khan, attacked Baghdad.

Sheik Iezid died in the fighting close to the Suleiman bridge. The caliph al-Mutasim was taken prisoner and beheaded by the Mongols. The city was sacked and cruelly leveled. The splendid city, which for five hundred years had been a center for arts and letters and the sciences, was reduced to a heap of ruins.

By good fortune, I was not on the scene of that crime perpetrated on civilization by these savage conquerors. Three years before, when the generous Prince Cluzir Shah died—May Allah give him peace!—I had gone to Constantinople with Telassim and Beremiz.

Every week I visit him, and, at times, I am envious of the happy state in which he lives with his wife and his three sons. When I see Telassim, I am reminded of the words of the poet:

*Sing, birds, your purest songs!*
*Shine, O Sun, with your sweetest light!*
*Loose your arrows, God of love!*
*Blessed be your love, my lady! Great your joy!*

There is no question that, of all problems, the one Beremiz solved best was that of life and love.

Here I end, without numbers or formulas, the story of the Man Who Counted.